Focus on Computer Graphics
Tutorials and Perspectives in Computer Graphics
Edited by W.T. Hewitt, R. Gnatz, and W. Hansmann

M. Grave Y. Le Lous
W. T. Hewitt (Eds.)

Visualization in Scientific Computing

With 121 Figures, 57 in Colour

Springer-Verlag
Berlin Heidelberg New York
London Paris Tokyo
Hong Kong Barcelona
Budapest

Focus on Computer Graphics

Edited by W. T. Hewitt, R. Gnatz, and W. Hansmann
for EUROGRAPHICS –
The European Association for Computer Graphics
P. O. Box 16, CH-1288 Aire-la-Ville, Switzerland

Volume Editors

Michel Grave
ONERA, DMI/CC
29 Avenue de la Division Leclerc
F-92322 Chatillon
France

Yvon Le Lous
EDF/DER, Service IMA
1 Avenue du Général de Gaulle
F-92141 Clamart Cedex
France

W. Terry Hewitt
Computer Graphics Unit
University of Manchester
Computing Building
Manchester M13 9PL
United Kingdom

ISBN-13: 978-3-642-77904-6 e-ISBN-13: 978-3-642-77902-2
DOI: 10.1007/ 978-3-642-77902-2

Library of Congress Cataloging-in-Publication Data.
Visualization in scientific computing / [edited] by M. Grave, Y. Le Lous, W. T. Hewitt. p. cm. – (Focus on computer graphics) (Eurographic seminars). Includes bibliographical references.
1. Computer graphics. 2. Supercomputers. 3. Visualization. I. Grave, M. (Michel), 1952-
II. Le Lous, Y. (Yvon). III. Hewitt, W. T. (W. Terry). IV. Series. V. Series: Eurographic seminars.
T385.V59 1994 502'.85'66–dc20 93-15289 CIP

Softcover reprint of the hardcover 1st edition 1994

Cover: Konzept & Design Künkel, Lopka GmbH, Ilvesheim, FRG
Typesetting: Camera-ready copy by authors/editors
SPIN 10053623 45/3140 – 5 4 3 2 1 0 – Printed on acid-free paper

Preface

Visualization in scientific computing is getting more and more attention from many people. Especially in relation with the fast increase of computing power, graphic tools are required in many cases for interpreting and presenting the results of various simulations, or for analyzing physical phenomena.

The Eurographics Working Group on Visualization in Scientific Computing has therefore organized a first workshop at Electricité de France (Clamart) in cooperation with ONERA (Chatillon). A wide range of papers were selected in order to cover most of the topics of interest for the members of the group, for this first edition, and 26 of them were presented in two days. Subsequently 18 papers were selected for this volume.

The presentations were organized in eight small sessions, in addition to discussions in small subgroups. The first two sessions were dedicated to the specific needs for visualization in computational sciences: the need for graphics support in large computing centres and high performance networks, needs of research and education in universities and academic centres, and the need for effective and efficient ways of integrating numerical computations or experimental data and graphics. Three of those papers are in Part I of this book.

The third session discussed the importance and difficulties of using standards in visualization software, and was related to the fourth session where some reference models and distributed graphics systems were discussed. Part II has five papers from these sessions.

The fourth session was dedicated to presentation of different application systems, and two of them form Part III.

Many papers were received on "rendering techniques", and this emphasized, if necessary, the importance of global visual representations in visualization graphics. Different methods for representing 2D or 3D scalar fields were presented, including several papers on so-called "volume rendering" techniques. In Part IV, six papers have been selected from the two sessions dedicated to this topic.

Finally, user-computer interactions were discussed, although they were present in most of the previous presentations, and two papers have been selected for Part V.

During the workshop, it appeared important to start the identification of different topics of interest for future work. After discussion, four subgroups were created, and had separate meetings, with a general synthesis at the end of the workshop:

- Visualization pipeline, distributed visualization and interaction, chaired by Georges Grinstein, from University of Lowell, MA, USA,
- Application environments, visualization tools and animation, chaired by Lesley Carpenter, from Numerical Algorithms Group Ltd, UK,

- Flow problems, chaired by Hans-Georg Pagendarm, from DLR Institute for Theoretical Fluid Dynamics, FRG,
- Presentation methods, chaired by Jarke J. Van Wijk, from Netherlands Energy Research Foundation.

Those four groups, despite the shortness of the workshop, provided interesting guidelines for subsequent activities of the working group, and others.

Michel Grave, Yvon Le Lous, Terry Hewitt

Contents

List of Colour Plates

Numbers in parentheses indicate the pages of reference.

Part I

General requirements

1 Scientific Visualization in a Supercomputer Network

U. Lang, H. Aichele, H. Pöhlmann, R. Rühle

ABSTRACT
Larger amounts of data produced on supercomputers have to be analysed using visualization techniques. As most of the users are not located at supercomputer sites, fast networks are needed to visualize computed results on the desk of the user. Centralized and distributed visualization modes and services, based on video equipment, framebuffers and workstations are discussed. Transfer rates for visualization purposes in local and wide area networks are derived. They are compared to transfer rates between supercomputers and workstations.

1.1 Introduction

Since November 1986 the University of Stuttgart Computer Center (RUS) has offered Cray-2 computer power to industrial and academic customers. Huge amounts of data are produced by calculations done on such machines. The appropriate way to analyse and understand these results is to visualize the data. This means, that graphical representations of computed results have to be presented to the user of the supercomputer. Supercomputers are usually run by Computer Centers located at centralized places, whereas users are distributed across a campus or whole countries. The amount of generated information makes it necessary to use fast networks to connect customer machines to the supercomputer. For visualization purposes a user should at least have access to a color workstation with 8 bitplanes, graphics hardware is sometimes recommended.

In addition to delivering raw computing power, RUS gives support in visualizing computed results by developing and distributing libraries and software tools and by offering access to specialized visualization equipment.

1.2 An environment for scientific visualization

The RUS Computer Configuration with its networks is shown in plate 1. The Cray-2 is connected to an UltraNet with 800 Mbit/s, to a Hyperchannel with 50 Mbit/s and to an Ethernet with 10 Mbit/s. Frontend machines from different vendors offer terminal access to the Cray-2 for pre-/postprocessing purposes and for file storage. User machines are usually attached to the Computer Center via Ethernet. The campus Ethernet is connected to BelWue (Baden-Württemberg Extended LAN) [6], which itself is part of the Internet. Additional high speed access paths can be made available via VBN (forerunner broadband network) [7]. Such a high speed access exists between the UltraNets of RUS and Alfred Wegener Institut in Bremerhaven, 800 km from Stuttgart (see¡ plate 2). Additional access possibilities via FDDI (100 Mbit/s) and ISDN (64 kbit/s) to the Cray-2 will be tested in pilot projects in the near future. The UltraNet also offers a framebuffer, which can be addressed from any machine on the UltraNet.

1.3 Visualization methods in a distributed environment

To better suite the visualization needs and to make the different visualization methods possible, RUS has requested an extension of the environment as depicted in plate 3.

Depending on the type of application, different working modes are possible in a distributed environment. Complex calculations needing hours of cpu time are usually executed in a batch mode, whereas calculations with appropriate time scale may be controled interactively. Amount and complexity of data defines, whether visual analysis of calculations may also be done interactively. The different working modes of scientific calculation and visualization will be presented in the following chapters.

1.3.1 Interactive realtime calculation

In this case the calculation on the supercomputer is fast enough to allow realtime interactive visualization and analysis of the ongoing calculation. The capability to control and steer the simulation offers possibilities otherwise not available. This greatly reduces the number of simulation runs and the turn around time compared to a batch mode of working. There are different ways to display data depending on their quality and quantity (see plate 4).

Transfer of pixel images

Data produced on the Cray-2 are converted to images during calculation and transfered directly to the UltraNet framebuffer. The scientist can see the progress of the calculation on the framebuffer screen and interactively steer the calculation. As the UltraNet framebuffer is a pure output device, keyboard and mouse of a workstation is used for feedback into the calculation.

A typical application suited for this type of visualization is in the area of computational fluid dynamics with dense grids (see cover page of [5]). If the meshes with variables to be displayed map into screen regions smaller than the size of a pixel, it is more economical to transfer pixel images from the supercomputer, than to transfer the higher number of variables to be displayed. This visualization method needs a high speed network. A transfer rate of 100 Mbit/s is suitable at reduced resolution and display rate, whereas 800 Mbit/s are needed for a full screen resolution and a display update rate of 25 images/s.

RUS is implementing a portable subroutine library to support this type of application on the Cray-2 and Convex.

Transfer of graphical objects

Graphical informations produced on the supercomputer are transfered to the workstation and transformed into images. The scientist can see the progress of the calculation on the workstation screen and interactively steer the calculation. Feedback is possible via keyboard, mouse or other input devices of the workstation.

The connection between the application code on the supercomputer and the displaying process on the workstation is established via task-to-task communication based on sockets. Sockets are communication endpoints of the TCP/IP protocol available on most computers with Unix operating system. Based on these sockets a subroutine library has been created to support a two way communication of data with conversion of internal data representations between the different machine types. To speed up network transfer buffering mechanisms are incorporated [2]. Graphical objects can be of different types.

Application Specific Objects Graphical objects may be application specific components of a system to be simulated. An example for a code defining object descriptions to be transfered via network is a slightly modified version of the molecular orbital program MOPAC. It calculates iteratively energy states for molecules. Transfered across the network are informations like type of atom, coordinates in 3D-space, bond information, etc. On this level of abstraction the amount of information being transfered across the network is minimized. On the workstation, the molecule is displayed and the coordinates of the atoms are updated online while the simulation continues on the supercomputer.

Based on TCP/IP sockets a graphics server for MOPAC was implemented. The application was demonstrated on DECWorld, Cannes 1988, with the Cray-2 being located in Stuttgart and the Workstation in Cannes. The connection was based on a leased 64 kbit/s line.

General Graphical Objects A more general approach to define graphical objects for distribution across networks is based on high level graphics libraries like PHIGS, Iris GL or Doré. In this case the distributed objects are the graphics primitives of the selected library. Two graphics libraries were implemented at RUS. One for Iris GL, the other for PHIGS. The PHIGS version was used for a presentation of an animated particle flow on DECWorld, Cannes 1988, whereas the Iris GL version was demonstrated on the inauguration of the VBN (140 Mbit/s) between Karlsruhe and Stuttgart [1]. Particle flow animation is possible starting at a transfer rate of approx. 300 kbit/s, but higher rates are desirable.

The main difference between the two cases is in the amount of information transfered across the network. By using high level object oriented graphical representations of results, the amount of network trafic can be dramaticaly reduced, thus gaining speed in the display of information. This approach is well supported, if the graphics software library on the workstation offers 3D objects as basic elements and methods for defining new objects.

Transfer of application data

This is the classical way of postprocessing results produced on the supercomputer. In this case the data are transfered to the workstation and rendered into images while the calculation is still going on. Once again the scientist can see the progress of the calculation on the workstation screen. These simulation codes were usually written for later postprocessing of data and now adapted to immediate display of results. Feedback and interactive steering of the calculation is possible if the simulation code was prepared for this. This method usually transfers the biggest amount of data.

The connection between the application code on the supercomputer and the displaying process on the workstation is again established via task-to-task communication based on TCP/IP sockets. To filter the data, map it into a graphical representation and render it into pixel images, all at the same time requires additional computing power on the workstation. In most cases it is necessary to have a superworkstation to visualize the behaviour in realtime.

1.3.2 Interactive realtime analysis of results

An interactive realtime simulation is not possible, if the computation done on a supercomputer takes hours to be performed. But an interactive analysis of the results with realtime behaviour is still desirable. Thus the human capabilities to see geometric and space/time relationships can still be used. An intermediate step is needed before results can be visualized.

1. Storage of Results

   Computations are done in batch mode on the supercomputer. Data are either kept on the supercomputer or transfered to a fast fileserver. Data kept on the supercomputer can be postprocessed there. Data transfered to the fileserver are passed on to the workstation for postprocessing. Thus the supercomputer needs not be involved in the storing and handling of results for postprocessing. An additional online output of some intermediate results can be useful to control the calculation.

2. Visualization of Results

   A transformation of the stored results into graphical animated representations can be done by generating pixel images, or by transfering graphical objects. The same explanation given for interactive realtime calculation applies here. Interactively it is possible to analyse the results in slow motion, single step or realtime, to zoom, pan or rotate the graphical representation, to alter display methods, colors or influence any other component of the image creation step.

1.3.3 Interactive realtime image analysis

It takes too long to get pixel image sequences with animated effects on screen, if the complexity of images to be produced is very high, or special, compute intensive image generation techniques like ray tracing are used. To get a display of results in realtime for the human analysis, images have to be produced as pixel data in a separate intermediate step. They are then either recorded with single framing methods on video tape recorders or video disks for later review, or stored on the fast fileserver to be reviewed using the UltraNet framebuffer.

1. Storage of Results

   Computations are done in batch mode on the supercomputer. An additional output of some intermediate results can be useful to control the calculation. Resulting data are transfered to the fast fileserver for later analysis, thus freeing the supercomputer for further calculations.

2. Image Generation

   The transformation of the stored results into graphical representations can be done in different ways.

   - Special graphics hardware on the superworkstation generates the images.
   - Ray tracing methods on the supercomputer or fast fileserver are used to generate images.

3. Display of Images

   The generated images can be displayed in animated sequences by using the UltraNet framebuffer or video equipment.

   - Usage of Video Equipment

     The images are transfered in a single framing method to the realtime videodisk. Due to the reduced resolution of video compared to a workstation display a

Text	4.8 - 9.6	kbit/s
Line Graphics	9.6 -19.2	kbit/s
Colored Graphics	0.5 - 3	Mbit/s
Simulation	16 - 64	Mbit/s
Animation	750 - 1000	Mbit/s

TABLE 1.1. Required Transfer Rates between a Workstation and the Cray-2

certain amount of information is lost. It is possible afterwards to analyse the results in realtime on a TV monitor. The display speed and direction can be altered interactively, thus allowing a detailed analysis of results (slow motion, single step or backward display). Different scenes can be compared by quickly switching between them or defining new scene sequences. The video motion pictures can be copied on video tapes and distributed to customers without highend workstations or visualization equipment.

- Usage of UltraNet Framebuffer

  Images stored on disks of the fast fileserver or Cray-2 can be displayed in animated sequences on the UltraNet framebuffer. In this case images still have a resolution of 1280*1024 pixels. In addition to the analysis methods using video equipment colors can be changed to give further insight into the results of the calculations. Via scan converter animated image sequences from the UltraNet framebuffer can be recorded on video tape without using the single framing technique.

  Software has been generated at RUS to read compressed sequences of pixel images from the disks of the Cray-2 expand the to full images and display them on the UltraNet framebuffer [3]. The first implementation reaches a transfer rate of 266 Mbit/s which gives a display rate of 25 images/s at a quarter resolution. Using the framebuffer hardware option to double the size of pixels in each direction still displays full screen images.

1.4 Network requirements

Different visualization methods have been explained. They require a wide range of transfer rates shown in table 1.1 and 1.2.

Simple alphanumeric control of ongoing calculations can be done based on normal terminal lines. The same applies for monochrome line drawings. Transfer rates up to 19.2 kbit/s are sufficent for this purpose.

Medium size (512*512) pixel images with a depth of 8 bit/pixel can be displayed every second at a transfer rate of 2 Mbit/s. This does not give an impression of animation, but control of simple animations may be possible. Using compression techniques on the pixel images and difference encoding of images sequences compression factors between 20 and 150 can be reached [4]. If supercomputer and workstation have enough cpu power to compress and uncompress the transfered pixel images, display rates of 20 images/s should be reachable. Necessary transfer rates of 1.1 Mbit/s have been measured on the campus Ethernet in Stuttgart using task-to-task communication between the Cray-2 and a workstation.

Method	Transfer rate Mbit/sec
ISDN (64 kbit/s)	0.056
Ethernet (10 Mbit/s)	
FTP (Disk - Disk)	0.3 - 0.6
Task - Task SGI Iris	1.1
NFS (Disk - Disk)	1.6
Memory - Memory (Sun)	1.5 - 3.5
Hyperchannel (50 Mbit/s)	
ftp	1.4 - 1.8
Memory - Memory	3
VME-bus (FEI 3)	
Memory - Memory (Sun)	30
UltraNet (800 Mbit/s)	
ftp (Sun)	4.5
Memory - Memory (Sun)	33
Framebuffer	747

TABLE 1.2. Measured Transfer Rates between a Workstation and the Cray-2

This technique is not applicable to the UltraNet framebuffer, because it can't handle compressed images.

For an impression of smooth animation at least 15 images/s have to be displayed. To transfer medium size images with a color palette of 256 entries a transfer rate of 30 Mbit/s is needed for uncompressed images.

A full size UltraNet framebuffer image has 1280*1024 pixels with 24 bits/pixel. This is approx. 31 Mbit/image. At a display rate of 25 images/s a transfer rate of 786 Mbit/s is needed. The measured transfer rate of 747 Mbit/s results in approx. 22 images/s. This gives the impression of smooth animation on the double buffered framebuffer.

1.5 References

[1] *Numerik-Labor Bundesrepublik*, 1988.

[2] Hartmut Aichele. Verteilte Grafik zwischen Supercomputern und Workstation. Technical report, Rechenzentrum Universität Stuttgart, 1990.

[3] Daniel Banek. Anwendung vektorisierter Kompressionsalgorithmen zur animierten Bilddarstellung in Hochgeschwindigkeitsnetzen. Technical Report Interner Bericht Nr. 45, Rechenzentrum Universität Stuttgart, 1990.

[4] W E Johnston, D E Hall, J Huang, M Rible, and D Robertson. Distributed Scientific Video Movie Making. Technical report, Advanced Development Group, Lawrence Berkeley Laboratory, University of California, USA.

[5] B H McCormick, T A DeFanti, and M D Brown. Special issue on Visualization in Scientific Computing. *Computer Graphics*, 21(6), November 1987.

[6] P Merdian. *Rechnernetze*, April 1989.

[7] R Rühle. Visualization of Cray-2 simulations via the 140 Mbit/s Forerunner Broadband Network of the Deutsche Bundespost. In *Proceedings of the 21st Semi-Annual Cray User Group (CUG) Meeting, Minneapolis, Minnesota, USA, April 25-29*, page 129, 1988.

2 Visualization Services in Large Scientific Computing Centres

Michel Grave and Yvon Le Lous

2.1 General

The R&D divisions of EDF and ONERA are two large research centres using supercomputers. They have many similarities in the architectures of their computing facilities and in the way they are operated. In this paper, we summarize common aspects of the management of visualization services in such environments.

In these centres, the overall architecture is based around supercomputers, with remote access from mainframes acting as front-ends, or from specific Unix workstations. In general, mainframes (CDC, IBM, ...) where installed before the arrival of the supercomputers, and remained as file servers, interactive servers for pre and post processing or for job preparation, or for I/O management like printing services or networking. All that constitutes an heterogeneous world, with a large variety of centralized equipments and terminals interconnected by local (Ethernet, NSC hyperchannel, Ultra, ...) and long distance networks (Transpac, X25, dedicated lines, ...), and with several operating systems, including Unix(es) and proprietary ones (NOS/VE, MVS, ...).

This architecture, based around mainframes, has evolved during the 80's, first by incorporating department computers, and then by adding more and more workstations. However, the centralized equipments still progress by a factor of 10 and more every decade. There is also a new tendency for the 90's, towards the interconnection of different sites together, mainly at the European level, for facilitating the cooperation between partners in European projects.

The computer's operation is also very similar from one centre to the other. Most of the work supercomputers is performed in batch mode, but there is today a growth in interactive access mainly allowed by the large increase of their main memories . The operations is usually well planed, and the balance between production and research work well organized, varying upon the load of the main computers.

Users are often distant, and are generally engineers whose main concern is the development and use of numerical codes. *Operation teams* are mainly concerned by the optimal use of the ressources (CPU, storage, printing, ..). Even if they intervene the choice and implementation of networks, they generally do not control the acquisition and management of terminals, workstations and department systems.

In this general context, the size and complexity of numerical simulations implies a growing volume of results, which can be analyzed only in a graphical way, in order to understand the physical phenomenons modeled. This is the domain of Scientific Visualization, and it is clear that the maximum amount of graphical services has to be provided to the users of large computing facilities.

The team in charge of the choices, developments and installation of these visualization tools is situated between the users and the operation team. The existing (and imposed) configuration has to be used as efficiently as possible, in the service of applications completely defined by the users. This team can be considered as the architect or integrator of graphic hardware and software, that have to be adapted as well as possible to the existing environment. From what has been described previously, it appears that the solution chosen by it will need some time to be diffused through the overall organization, and that this inertia limits the speed at which computer graphics can be widely introduced.

In this paper, we first present the general needs and behaviour of the users We then detail different steps and resources necessary in the visualization works, and give an overview of the general solutions that can be adopted in large scientific computing centres.

2.2 Needs and behaviours of users

From a numerical simulation code design, to its use in the solving of practical cases, the user goes through different phases in its work. During each of them, its needs visualization tools can be different. We can consider that there are three types of tools:

Standard (or general) tools , are the software and hardware widely diffused, available for all users of a computing centre, and for which have been organized a standard documentation, training, support and assistance. They can be characterized as reliable, simple, available, and stable in time. They also very often require a minimal personal investment from the user. They are part of the basic computing culture of them, on the same level as operating systems, programming languages or file management systems. In addition to the normalized kernel systems (GKS, PHIGS), more adapted for graphics packages development, different application packages can be found. NCAR Graphics, UNIRAS, DISSPLA, MOVIE.BYU are among the most frequently encountered, and a user of one of these has good chances to find them again when he moves from one centre to another. For graphic hardware, Tektronix, Versatec, and now Silicon Graphics are among the most currently found.

Specialized (or specific) tools are not necessary available to all users, and are not necessary for a wide range of applications. There are necessary for solving a problem or a class of specific problems. They usually require more technical support from the Visualization team, and require some personal investment from the user. He can sometimes accept some unreliability and unavailability. These tools can be prototypes. They are often developped during interactive exchanges between users and developers, and helps them to better understand visualization problems, and design new tools that will be made more widely available later. Systems for volume rendering (like TAAC-l on Sun or PIXAR), "realistic" display (RenderMan, Oasis, ...), or interactive flow visualization (PLOT3D, GAS, MPGS, ...) for example fall into this category.

Communication tools serve in the exchange of information with other people from the same centre or from another one. Ranging from simple tools - like screen hardcopies - to sophisticated ones - like 35mm films with sound -, these tools will need a wide variety of equipments and support that will not necessary be in the computer centre environment itself. Film or video post-production is one of the examples. For this communication, the quality of the result will always be a criterion more important that the ease to produce it. They are the most visible parts of the work done by a team in scientific computing, and the quality of these communication tools can have an important impact on the fame of a centre.

The need for such tools evolves for during the different phases of the user's work. We can roughly consider 4 phases: debugging, preparation, production and synthesis (or communication).

During the *debugging phase* algorithms and procedures are studied and validated. Graphic tools are then used to visualize the behaviour of some parts of the programs, after or during execution. Sophisticated "debugging" techniques can be programmed, by adding for example temporarily visualization commands in some parts of the program. Pre-processing tools are quite not used since "standard" test cases are taken to validate the codes. In all cases, the user is interested much more by the correctness and efficiency of its codes, and does not require much sophistication for the visualization. He uses simple, reliable graphic tools, and usually standard ones. In some cases however, some specific tools can be involved for "monitoring" jobs.

During the *preparation phase* the user prepares its data, and the formats that will be used for visualizing later the simulation results. What is needed is interactive and standard tools. At this stage, a rather small number of graphics is produced, and their quality does not need to be too high.

The next step, *production*, is the real step of numerical simulation. Many runs are performed either to verify that the hypothesis are physically acceptable, or to really analyze some phenomenon. This phase can be repeated several times, for refining results or test new hypothesis. Many graphics are produced, and it is generally useful to have a software running in batch mode, but compatible with an interactive version used in preparation. However, since the number of graphics is large, interactive tools can be needed to quickly scan them. The quality of images is here again not as important as their speed of production. For example, animation can be done in video, but within a few hours, or black and white laser prints are acceptable if hundreds of images can be produced. Usually, the graphic tools used during this phase will still be mainly standard, except when the type of code or size of problem is new and requires more specific tools.

The last step, *synthesis* or *communication*, is where the most significant results are selected, and visualized using sophisticated tools. Still images or films, for papers, conferences or general promotion are produced with high quality systems. In some cases, the advanced techniques developped for that purpose are afterwards refined, and lead to the implementation of new tools that can become "standards" some years later.

It appears that there is a great variety, in the quality and complexity of tools to be provided, on the quantities of graphics to be handled, on the interactivity, and on the user's support necessary. Those different factors have an effect on the nature of hardware and software to handle, and on the human resources to provide.

Since in a large centre, projects often include all the phases previously defined, computer environments can be very diversified. An engineer can work on a workstation or a supercomputer, he can as well use a small hardcopy device in the preparation steps, as a large graphic printer producing hundreds of pages in production steps. In the same way, for the quantities of data handled, local capacities of workstations will be enough during preparation, but large file servers will be needed during production. It is then important to provide similar environments to a user moving from one system to another, and then to provide him portable software when possible.

2.3 The different steps of the visualization process

The visualization process, consisting in the transformation of numerical data from experimentations or simulation into visible information can be qualified using different criterions:

- nature of data processing required.
- nature and amount of computing resources needed.
- bandwidth and volume of data exchanges.

We consider that this process can be subdivided into 3 steps:

Interpretation is the step where simulation (or experimental) data are transformed into data that can be graphically represented (usually colored geometrical entities). For example a temperature field is transformed into isocurves or isosurfaces, a velocity field into a set of arrows or particle traces, etc ...

This step usually requires mainly computing power and memory for handling data. The level of interactivity is usually low, and the amount of data produced can sometime much smaller than the original one.

Visualization is where geometrical data are transformed into graphical primitives. This is where we find the usual basic graphic libraries like GKS or PHIGS.

This step still requires more computing resources than graphical ones. The level of interactivity is higher than in the interpretation step, and the amount of data produced reduced again because everything is not necessarily "visualized";

Display is the step where primitives are effectively transformed into visible objects like pixels. This is the level of device drivers.

Graphical resources are critical at this level, and high levels or interactivity are usually required.

It is clear that the border between visualization and display varies very much upon the type of hardware and software used. For example, on a 3D workstation, driven by a 3D library, a 3D transformation will be performed at the display level, but if it is driven by a 2D library, it will be done at the visualization level.

From data to pixels, these 3 steps are usually sequential, even if the graphic package does not identify them precisely. They are not totally independent, and a specific interpretation can be designed in view of a specific graphical representation.

This splitting of graphic systems into 3 parts offers big advantages for the portability and adaptation to a specific hardware configuration, but it also provides flexibility during visual analysis of results:

- By simply acting on the display parameters, it is possible to quickly modify some attributes of the displayed image (color table manipulation, zoom, ..) or even 3D geometrical transformations sometimes.
- By acting on the visualization parameters, it is possible to modify the set of graphical primitives handled, and their attributes and environment.

2.4 Solutions

2.4.1 Standard tools

A Common culture for the users of a scientific computing centre, in order to garanty the perenniality of developments, but allowing the evolution of the architecture, can only be achieved by using some standards for:

- graphic software
- data exchange
- hardware and operating systems

- networking

In the field of standard tools, services provided by the visualization team must be clearly defined, and usually include:

- An up to date catalog of available tools.
- Training and documentation, taking into account that they have to be accessible with a minimal personal investment, in order to be used by non-computer people and students.
- User's support
- Site licensing negotiations.
- Interface with suppliers.
- Animation of internal user's groups and participation to external groups and normalization committees.

In the following, the word "standard" must be taken in its general meaning, and not restricted to what is defined by ISO, ANSI, AFNOR or any other official institution. In fact there is always a big gap between the official standards and the real market's needs, and "de-facto" standards need often to be used, eventually for a limited period before the definition of an official one. They can be for example proposed by some manufacturers whose wide diffusion forces its competitors to be compatible, or by a group of manufacturers. In the case of application software, on the other hand, some packages become widely used only because they have not much competitors.

Graphic packages

For basic libraries, GKS for 2D, and PHIGS for 3D are presently two international official standards. There is a 3D extension to GKS (GKS3D) but it suffers from the comparison with PHIGS, which has more functionalities, but above all is much more supported by the international community and by manufacturers.

In a large centre, the problems linked to the introduction of GKS and PHIGS are:

- Functionalities are often lower than those provided by already used application packages, which implies the development of additional levels on top of the basic package.
- Choices have to be made between products from different suppliers, with different levels of quality in reliability, portability, conformance to the official standard, and maintenance. This implies long evaluation procedures on different systems. On a given equipment, proprietary implementations should be privileged, since it usually means good performances, but this can leed to problems in portability and compatibility between sites (GKS metafiles are a good example of it as well as graPHIGS from IBM).
- Some "de-facto" standards are sometimes difficult to leave (PlotlO IGL from Tektronix or GL2 from Silicon Graphics for example).
- Performances are sometimes poor and imply the use of specific graphic accelerators.

At the time of writing an extension to PHIGS (PHIGSPLUS) incorporating shading models and high-level primitives is under study, but many versions of it are already provided by workstations manufacturers; Moreover, implementations of PEX, which is the integration of PHIGSPLUS capabilities into the X-Window system, are expected for the beginning of 1991. The adoption of PHIGS in the middle of 1990 would then not be an up-to-date choice, and this has to be examined carefully, since as said previously, in large centres, it takes time to introduce something, and choices remain for a long time.

In the field of graphical application software, there are usually many local and specific packages, developed generally by the users themselves, but the number of widely diffused and well supported ones is very small. UNIRAS and DISSPLA are the most famous. MOVIE.BYU and NCAR Graphics (rewritten on top of GKS) are also often encountered but have a very poor support. Usually, the maintenance and assistance to users for these packages require some human resources from the visualization team. This small number of packages, and their maturity (in general they are more than 10 years old) has the advantage of making them kind of "de-facto" standards, since they are available on many systems and have lots of device drivers. In practice, an experienced user arriving from another centre generally knows one of them.

Data exchange formats

It is very important to have standard formats for the exchange of information:

- Between numerical simulation systems and visualization ones, so that different graphic packages can be used to analyze a result.
- Between visualization systems, to use complementary functionalities of them.

For the first category, there is no really universal standard, even if some formats are often used, like CDF (Common Data Format) from NSSDC, HDF (Hierarchical Data Format) from NCSA, or the very simple "MOVIE.BYU" format, and others from software products suppliers (NASTRAN, PATRAN, ...). Usually file format converters are then needed, and many of them exist on a site.

For exchanges between graphic systems, CGM (Computer Graphics Metafile) is the international official standard, and is more and more encountered. However, it is presently only 2D, and very often CGM interpreters can accept only subsets of it. In practice, the interfacing of two different packages through CGM is not often easy, and requires some specific adaptation. GKSM (GKS Metafile) is not really a standard, and is very much related to GKS. There is always a need to write a transcoder for the GKSM of a manufacturer into the GKSM of another one, even if it is often rather simple to implement. PostScript appears sometimes as a graphical metafile format, since many interpreters for it exist, either built in hardware or software.

For images, a compaction method (usually Run Length Encoding) is often used, and different formats and transcoders already exist (Utah-RLE, TIFF, ...), but the official work of normalization in that field is only at its beginning.

Hardware and operating systems

Beside the classical terminals connected via serial lines to main computers (typically Tektronix or IBM), three type of equipments are standard components of the environment of a scientific computing centre.

- Unix based workstations, whose catalog still grows among manufacturers, like Sun, DEC, HP, IBM,... Power delivered by the CPU's are now measured in tens of Mips or Mflops. The graphic capabilities have also grown quickly, and hundred thousand 3D vectors drawn per second are now standard figures. Only the storage capacities of these systems, and transfer rates seem to be today a little bit weak. Even if the different versions of Unix and "Shells" can be sometimes confusing, they do not imply too much compatibility problems today.

- X-Window terminals are arriving strongly presently, and their functionalities, combined with their low cost, make them very attractive. However, some basic diskless workstations could offer good alternatives to them.

- PC and Macintosh are also frequently used in these configurations. "Mac's", with their well designed user interface, and the large applications catalog they provide make them very popular. PC's with X servers could also be an interesting alternative to X terminals, in some cases.

Networks

In addition to the local high-bandwidth networks connecting central systems (NSC hyperchannel or UltraNet for example), proprietary networks, like SNA-IBM, and TCP/IP on local Ethernet networks constitute the general skeleton for communication.

To allow the sharing of storage media by many different users, elaborate architectures, using gateways, IP routers, ..., need to be implemented. They quickly become very complex, and require rigorous administration.

In addition to traditional services like ftp, telnet, rcp and others, there are different tools available for building distributed applications (rpc, nfs, sockets, ...). If nfs is transparent for users and programmers, the other tools are not always easy to handle, and higher level layers are required (OSF/Motif, SQL-Net, ...)

For distributed applications, two major directions are emerging:

- Applications on workstations, using rpc to access Crays for CPU intensive tasks, and NFS or SQL to access data.

- Applications on Cray, with users interfaces on workstations, by the use of X-Window, and soon PEX.

In both cases, the user sees only the workstation, the requests to computing or data servers being transparent.

2.4.2 Specific tools

The complexity of the problems to solve, and the growing quantities of results to analyze can sometimes make standard systems not adapted, and justify the implementation of specific tools. Such tools, that always come in addition to standard ones are specific in the sense that they are applied to solve a specific problem or class of problems. They require more personal involvement from the users, and a close cooperation between them and the visualization team. Among such tools are:

- Animation systems

- Graphic superworkstations

- Image processing and analysis systems

- very high bandwidth networks.

Animation systems

When dynamics (function of time or of another parameter) is necessary for understanding simulation results, two kinds of tools can be provided. The first one is real-time (or pseudo real-time) animation on specific systems, when available, and the second one is frame by frame animation.

The recording of an image can take from a few seconds to several minutes, with direct or sequential access to the medium (erasable or not). Restitution can be done in real-time, within a time frame from one hour to several days, and frame by frame analysis or viewing speed modification is often useful.

- Film medium (either 16 or 35mm) is not very often used, because of the long recording and processing time required, and of the complexity of viewing equipments. It is mainly used for very final versions of animations.

- Frame by frame recording on Video Tape Recoders is becoming more and more widely used, and their low resolution is not a too big handicap. A minute of animation can be obtained within a few hours, and restitution tools are cheap and easy to manipulate. The different standards (PAL, NTSC,..) (BVU, BETACAM,...) (U-MATIC, VHS,...) can however be sources of problems for exchanges.

Graphics superworkstations

Offering many Mips, Mflops and 3D vectors per second, a minisupercomputer or a graphics superworkstation can become a system dedicated to visualization.

The power supplied allows real-time animation for large models, use of high-level interpretation and rendering techniques, and video recording can be done in real-time. There is also a growing number of application packages available on them, and they quite all offer an implementation of PHIGSPLUS.

Links with supercomputers must be high-bandwidth ones.

Image processing and analysis

Having been used for many years in medical imaging or seismic interpretation, 3D image processing an analysis systems begin to enter other fields. The basic problem is the exploration of arrays of "voxels", or more generally, 3D meshes, with scalar or vector data associated to each node. In addition to the classical filtering, thresholding or transformation techniques, slicing algorithms and isosurfaces computations are the bases of these systems. In many cases however, the geometry of the meshes is much more complex than in the first application fields, and many progress are awaited in new algorithms studies. PIXAR image computer, or Sun/TAAC-1 with their respective software are well known commercially available systems.

Very high-bandwidth local networks

The 2/3 megabits of ethernet-TCP/IP will always represent bottlenecks in the visualization process, when several computers are involved. Local networks with very high bandwidth are now available, with usual protocols.

- FDDI at 100 Mbits/sec, used to interconnect computers or ethernet networks.

- Ultranet, at 1 Gbit/sec for interconnecting mainframes and supercomputers, with very soon gateways with FDDI and Ethernet. On this network, we should also mention the existence of a frame-buffer, allowing the display of high-resolution raster images in real-time.

2.4.3 Communication tools

Beyond informal communication between users, within normal working relationship, and using tools defined previously, it is necessary to formalize at some times the communication process of scientific results.This is the case:

- For presentations.
- For publications.

If the quality of used media is always very important, the information content may differ according to the receiver of the message. The receiver may be roughly classified in the 3 following categories:

- specialists of the field, in charge of giving an advice about the method or the results.
- administrative or financial hierarchical authorities, in the company or at a higher level, who need a justification of allocated financial resources, or a presentation of the future resources needed.
- enlarged audience, from education and scientific magazines, to commercial TV and newspapers.

It is obvious that graphics play a fundamental role in such communication processes, and it is necessary to point out that tools and techniques used may be quite different from those used in simulation running phases and in the analysis of results.

Presentations

If computer graphics were born in the scientific laboratories where data were turned into meaningful trend-charts and diagrams, the business community discovered that graphics were a very powerful presentation tool and is always using it intensively in that way. Presentation graphics can be differentiated from scientific data analysis graphics in the following points:

- Information content is the result of a synthesis, and is presented in an enhanced fashion.
- in order to communicate a message, graphic representation may be adapted to the audience and the style of the presenter.
- to reach a high visual impact, artistic techniques may be used.
- different media can be used for presentation: analogical or solid graphic media like overhead transparencies, slides, movies and videotapes, digital media like floppy disks or cassettes and now optical media like CD-ROM.

If standard graphic packages can be used for making overhead transparencies and color slides on a thermal printer or film recorder, the help of a communication graphics specialist may be needed to graphically present the information in the best manner.

For videotapes, titles, charts and comments have to be added to scientific graphic sequences. For titles and charts, animation packages running on PC's or workstations are available (Freelance from LOTUS, Videoworks from Macromind, Wavefront's software,...). Editing the videotape, and adding sound effects may be made in a specialized video laboratory. The need of such a laboratory within the computing centre may be justified, depending on the number of video sequences edited every month. For 35mm movies, the problems are identical in nature, but resources and qualification needed are much higher, so that external professional are to be called. Collaboration between visualization support and scientific movie makers is very important for obtaining high quality results. Mention that in a near future, high definition television is likely to replace 35mm movies in Scientific areas.

Publications

Publications with large distribution like books and magazines are relevant of professional publishing techniques: high quality graphics on an external medium -paper or photographic film- may be given to the publishers. Note that color is still expensive and that most most color printer material require that four color separation, approximatively screened, be supplied to a printer.

For internal publications, electronic publishing packages are often used and scientific graphics and images are to be imported using the standards mentioned previously. Among the most widely used tools in the scientific community, we can mention TEX, which is more and more often accepted by publishers.

2.5 Conclusion

In a large scale scientific computing centre, it is essential to set up a graphics support team. This team is in charge of developments, advices and assistance in all stages of digital computing involving graphics and particularly visualization methods.

Between data processing centre people — essentially concerned by the monitoring of mainframes, supercomputers and networks — and scientists -essentially concerned with the development of numerical algorithms and with the solving of physical problems- visualization support people can be seen as computing architects in charge of integrating hardware and software graphic resources in a given environment.

They have to provide the basic graphic tools defining the standard level common to all users of the computing centre, and they have to collaborate with scientists for solving specific problems with specific visualization tools. Finally they have to know, outside the computing centre, where to find specialists able to help them to elaborate high quality communication graphics.

3 The Visualisation of Numerical Computation

Lesley Carpenter

ABSTRACT
Parallel processing has now made tractable the numerical solution of large complex mathematical models which are derived from across a whole spectrum of disciplines. Such powerful processing capabilities usually result in a vast amount of data being produced. However full value for this advance can only be realised if the engineer has an effective means of visualising the solution obtained. We need to develop efficient and effective ways of integrating numerical computation and graphical techniques.

3.1 Introductory remarks

Current practice in mathematical modelling usually has the scientist or engineer adopting a two-step approach: firstly computing the complete numerical solution and subsequently producing a graphical representation of the results. Any experiments that the scientist wishes to conduct (say, by changing a parameter of the mathematical model, or by adjusting the error control in the computation) require repetition of the whole process and thus there is no chance to experiment as the computation proceeds. The ability to solve large problems interactively is now available, however the present style of working still tends to be based on the 'batch oriented' model of compute and then plot.

As the size of the problem and the quantity of data increase, the limitations of this 'traditional approach' become increasingly apparent. In order to solve problems in a truly interactive mode, i.e., allowing the user to interact with his 'data' there is a requirement for significant computation and graphics power. Although there have been major advances in hardware (e.g., super-computers, graphical hardware etc.) in recent years, to date there has been little software available to utilise these. The landmark issue of ACM SIGGRAPH Computer Graphics – 'Visualisation in Scientific Computing', published in November 1987[2] identified the main aim of scientific visualisation as seeking to provide the scientist with the key to new insight through the use of visual methods. To date a good deal of software has been developed by research institutes (particularly in the U.S and largely as a direct response to this report) but little is actually available to the scientist or engineer for use in their own domain.

3.2 A model for visualising computational processes

Often computational modelling problems involve a design activity that is iterative, complex and computationally intensive; iterative in that the solution cannot be uniquely specified from a single parameter specification and computationally intensive because of the large volumes of data or complexity of the model, for example multi-dimensional systems. If we consider the steps which a scientist makes in order to develop and execute a mathematical model, we can begin to define a generic computational or reference model[5]; see figures 3.1 and 3.2

The first step is to define the mathematical formulation of the problem and select appropriate algorithms to perform the computation; this may be achieved via the use of graphical and or symbolic tools, possibly with the assistance of a knowledge based front end or an intelligent help facility to provide guidance on the selection of algorithms. This,

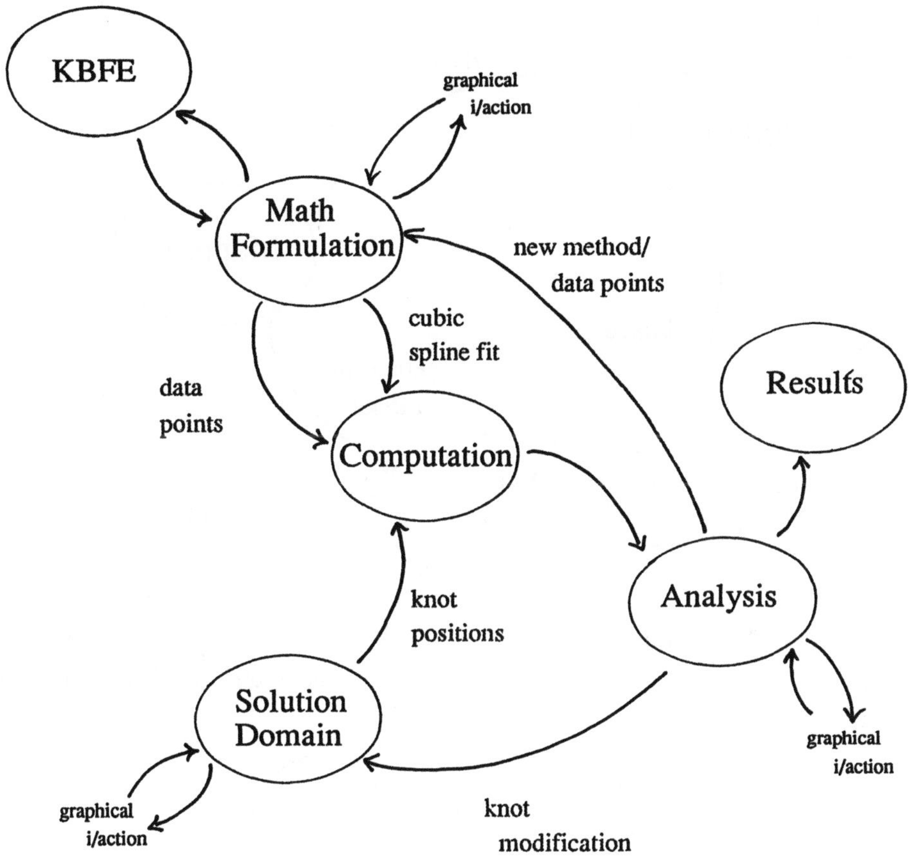

FIGURE 3.1. Computational cycle

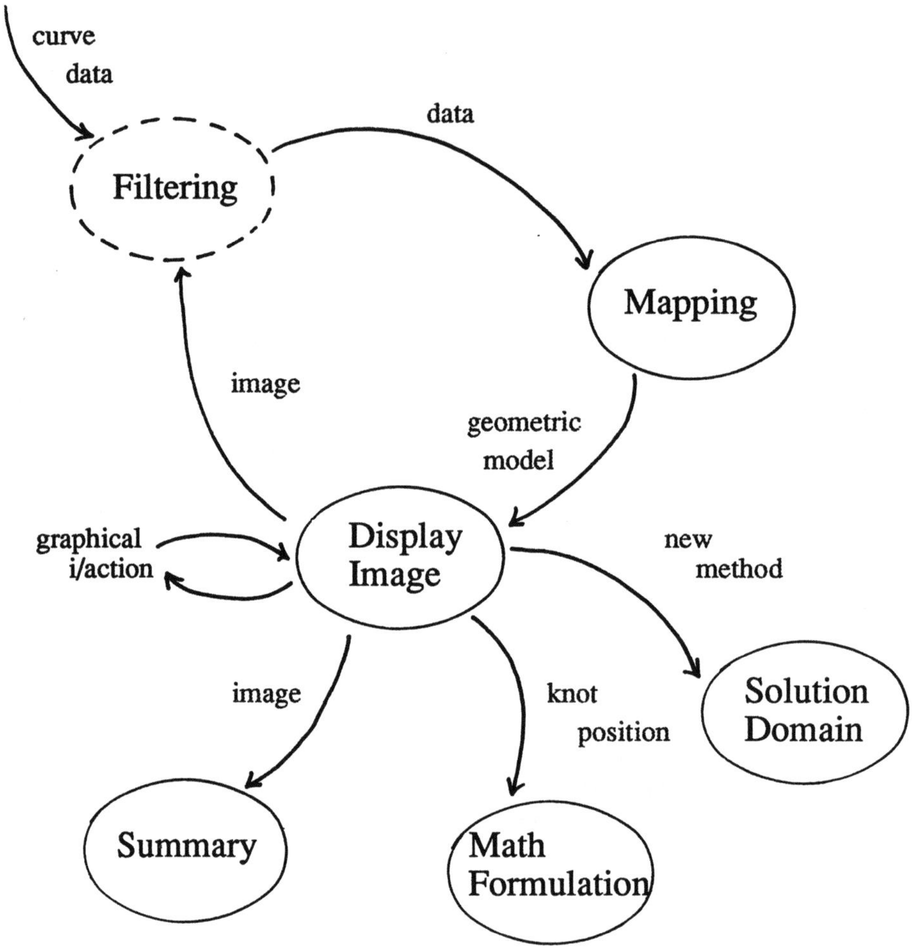

FIGURE 3.2. Analysis

coupled with the definition of the solution domain (e.g., definition of range of parameters on which the mathematical definition applies), provides the problem description to be solved using numerical computation. The analysis phase may either be invoked at the end of, or during, the computation, perhaps utilising graphical images in a monitoring capacity. Depending on the interpretation of the results the computational cycle may be re-invoked at any point, for example if the analysis reveals a problem with the numerical approximation then it may be necessary to select an alternative algorithm, a problem with the structure of a finite element mesh may involve re-assessing the solution domain etc.

The key element is that the scientist be allowed to formulate the model, select the style of graphical displays used, decide when to interrupt the calculation, etc., but that the system effectively remains in control by only allowing meaningful choices to be made. A complete interactive system developed to study the behaviour of complex Ordinary Differential Equation (ODE) solutions is described in[3]. The techniques described, although simple in terms of the computer graphics and interaction offered, show how effective visual tools can be in gleaning new insights in a specific field, representing and conveying result information and providing scientists with a totally new way of approaching their problems.

3.3 Data structure visualisation and algorithm animation

According to Tuchman and Berry two different, but related, fields of numerical visualisation can be distinguished; namely data structure visualisation and algorithm animation [4]. Algorithm animation attempts to model the process of the algorithm and can be used to visualise behaviour and performance through the use of images to represent aspects of an algorithm's execution. Data structure visualisation is, as the name implies, concerned with 'viewing' the contents of data structures, for example matrices. Data structure displays are becoming invaluable in aiding the understanding of the apparantly chaotic behaviour of some algorithms.

A relatively simple, yet effective use of algorithm animation is illustrated by the work undertaken by Hopkins in the development of the PONS system [1]; a collaborative effort between the University of Kent, NAG Ltd. and the IBM Bergen Scientific Centre. The first prototype version of PONS was developed as a direct response to user queries regarding the use of the NAG spline fitting techniques to fit smooth curves to a set of data points. A major problem faced by the user is generally the question of where to position the knots to obtain a realistic curve (users will often have an idea of what they expect the final approximation to look like). By using PONS with its straightforward user interface (see figure 3.3) users were able to experiment easily with various knot positions and obtain their final fits relatively quickly. The system also had the beneficial 'side-effect' of providing an effective teaching aid.

The effectiveness of data structure visualisation is illustrated in [4]. By developing visualisation tools for use in the field of numerical linear algebra research, Tuchman and Berry have been able to show how static and dynamic structures associated with matrix manipulation can be represented visually and subsequently be used to aid in the development of hybrid parallel algorithms for the singular value decomposition. The static structure is illustrated by the pattern of zero or nonzero elements (differentiated by colour or grey scales), whilst a combination of colour, highlighting and animation is used to reveal the active portions of the matrix.

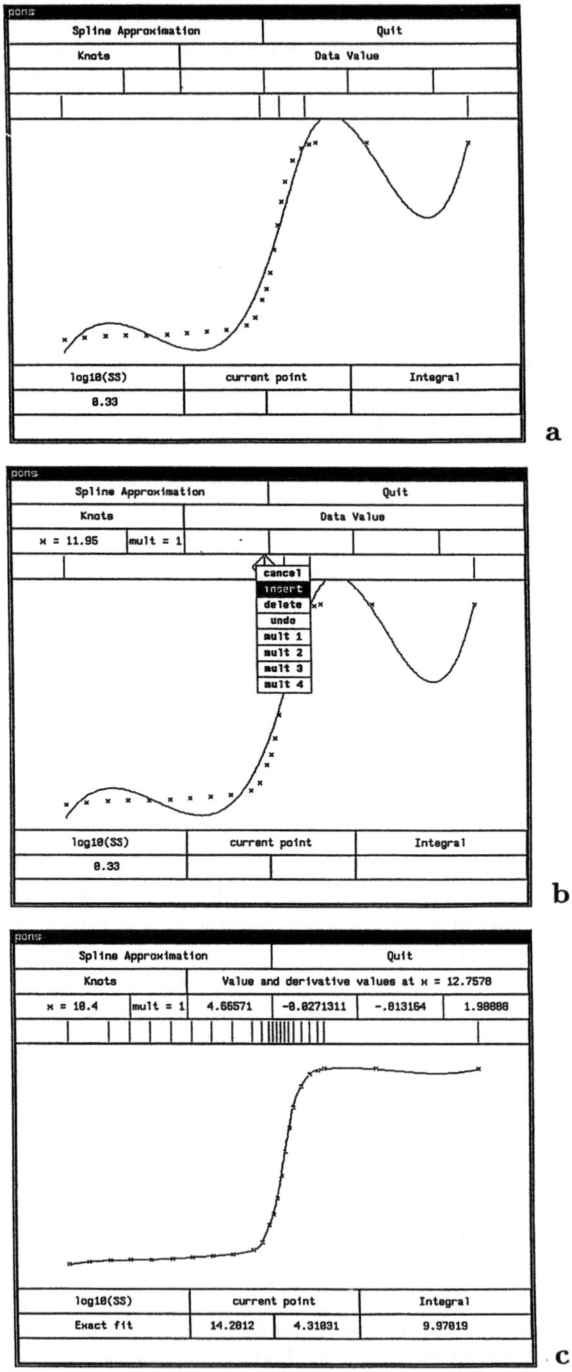

FIGURE 3.3. PONS and its user interface

3.4 Consideration of the target environment

We need to carefully consider the target environment which we believe will be available in the 1990's and upon which any visualisation software will be required to run since this will directly affect the type of tools, packages and software developed. Considerations include:

- Workstations and window based environments

  Support for a window based environment must be considered mandatory if we assume that powerful, graphical workstations are likely to be the target user machines. Strong consideration must given to the use of the X window system protocol, OSF/Motif, Open Look etc.

- Networking

  We should ensure that full use is made of the potential which networking can offer. The power, or specific capabilities, offered by individual distributed machines can be harnessed relatively easily; for example, the numerical computation being performed on a super-computer whilst all user interaction is via the graphical display facilities offered by a workstation.

- Parallelism

  The exploitation of parallel hardware architectures will be essential to ensure that the scientist can work freely without the traditional wait for either a computation to complete or an image to be displayed. We need to determine at what points in the pipeline it is appropriate to invoke parallel techniques; possibly in the numeric components and rendering but less so in the visualisation software itself?

- Standards

  The utilisation of de-facto and potential standards should lead to the development of a generic solution to the integration of numerics and graphics.

- Graphics base

  There is no clear view as to what a suitable, portable, graphics base should be. A number of major hardware vendors do now offer PHIGS/PHIGS PLUS as standard on their machines and as PHIGS is an International Standard we must seriously consider its adoption. However PHIGS, like the other graphics standards was designed for a serial, non windowing, computing environment and as such it may not prove to be appropriate. Likewise PHIGS was developed with the display of 3D hierarchical objects in mind rather than the broader remit of visualisation. The alternative is to adopt a proprietary software package (e.g., Dorè, AVS, Visedge, PV-Wave) or invest heavily in resources to develop a suitable base.

We must ensure that both vector and raster graphics can be supported; it may not be necessary to develop 'flashy' graphics but whatever is supplied must be flexible, extensible and useful.

3.5 Developing visualisation software

Until recently the scientist has been largely restricted to the use of generalised subroutine libraries (e.g., NAG Graphics Library, GKS, GINO-F etc.) or application specific packages (e.g., Finite Element packages) in order to perform any kind of visual interrogation of result data. Both categories have major disadvantages; for example, in the former case, there is a lack of extensibility and ease-of-use and in the latter case, it is difficult to transfer data between packages etc. Ideally, what the user requires is to have access to a suite of tools which:

- Minimise programming;
- allow applications to be tailored to individual needs;
- integrate with existing software packages;
- allow the integration of visualisation and computation components;
- are portable;
- take advantage of the distribution of the numerical, visualisation and rendering components in both parallel and non-parallel environments; and
- be suitable for use by differing types of user e.g., researcher and production engineer.

In order to achieve these aims a number of things are required:

- Development of a reference model for scientific visualisation and its subsequent consideration by relevant standards bodies;
- construction of a flexible toolset which allows interaction between the scientist and the
  - construction of their model
  - control of the numerical solution
  - style and composition of associated graphical images;
- development of portable, high-level graphics software which addresses the display of multi-dimensional data and which can be used in a parallel environment;
- research into the impact of visualisation requirements on the design and construction of numerical algorithms; and
- efficient utilisation of data management systems.

3.6 The GRASPARC project

GRASPARC is a project resulting from the United Kindom, Department of Trade and Industry Information Engineering/Advanced Technology Programme, Second Call—Parallel and Novel Architectures, Exploitation of Parallelism. The partners in the project are the Numerical Algorithms Group Ltd. (Project co-ordinator), The University of Leeds and Quintek Ltd.

GRASPARC proposes a model in which numerical solution and visualisation techniques are combined in an integrated procedure, thereby allowing the scientist to monitor calculations and adopt appropriate solution strategies as the computation proceeds. A key element is that the user should remain in full control of the navigation of his data throughout the system. A major objective of GRASPARC is to improve the interaction between the scientist and the parallel computer through the development of interactive visualisation software. The project plans to investigate, and where appropriate adopt, standards such as PHIGS PLUS and X-Windows as vehicles for the production of portable visualisation software. The exploitation of parallelism is considered to be essential to ensure that the scientist can work freely without the 'traditional wait' for either a computation to complete or an image to be displayed.

Work on the GRASPARC project is due to commence in September 1990.

3.7 Concluding remarks

We have highlighted the fact that many of the elements needed to undertake the solution of large, numerically intensive problems in an interactive environment are now at our disposal. As a result of hardware advances in recent years, the necessary computational power, memory and storage requirements etc. are becoming more readily available for use by the scientist or engineer in their place of work. Unfortunately, the development of software capable of exploiting the full potential of these advances is only just beginning to be addressed.

We have identified a requirement to develop software which is efficient, comprehensive and adaptable to the user. This adaptability will permit the user to impose their own style of working on the system rather than force him/her to work under the constraints usually associated with more traditional systems. Projects such as GRASPARC will provide valuable insight (with demonstrable solutions) into the design and development of generic visualisation software methodology and tools.

3.8 References

[1] T R Hopkins. NAG Spline Fitting Routines on a Graphics Workstation - the Story so far. Technical report, University of Kent Computing Laboratory, 1990. (due to be published September 1990).

[2] B H McCormick, T A Defanti, and M D Brown. Visualisation in Scientific Computing. *Computer Graphics*, 21(6), November 1987. special issue.

[3] F Richard. Graphical Analysis of complex O.D.E Solutions. *Computer Graphics Forum*, 6(4):335–341, December 1987.

[4] A M Tuchman and M W Berry. Matrix Visualisation in the Design of Numerical Algorithms. *ORSA Journal on Computing*, 2(1), 1990.

[5] C Upson and et al. The Application Visualisation System: A Computational Environment for Scientific Visualisation. *IEEE Computer Graphics and Applications*, pages 30–41, July 1989.

Part II

Formal Models, Standards and Distributed Graphics

4 Performance Evaluation of Portable Graphics Software and Hardware for Scientific Visualization

Nancy Hitschfeld, Dölf Aemmer, Peter Lamb, Hanspeter Wacht

ABSTRACT
In this paper we present an evaluation of the Programmer's Hierarchical Interactive Graphics System (PHIGS) [2, 3] as portable scientific visualization graphics software, for six different graphics workstations.

4.1 Introduction

For scientific visualization, the evaluation of the graphics software and the machine where it runs must be performed carefully to determine what tools are efficient for the graphical analysis of scientific problems. A comparison of graphics application performance across different hardware platforms is desired. Obviously, the measurement of hardware performance and the performance of a graphics package will give differing results.

Our group works in semiconductor simulation [6], which includes the numerical solution of semiconductor devices such as transistors, diodes, and sensors. We obtain as result, for example, electric potential, electron concentration, and carrier and current density functions. Normally, the simulations are in 3D and are functions of time; the traditional "x versus y" and "contour" plots are not sufficient for the visualization of such large result sets. The user of such a simulation program can only analyze his data using 3D graphics software on fast graphics hardware. In our applications, we generally need to draw 5000–20000 polygons with 10–1000 pixels per polygon for each frame. A drawing speed of approximately 10000 figures per second is necessary in order to be able to analyze the results in reasonable time.

For this reason, we are interested in the evaluation of different kinds of workstations, where the most important criterion is the graphics performance, particularly of portable graphics software such as PHIGS and Doré [5, 4]. In the literature, we have not found graphics benchmarks at this level. Normally, the evaluation of the graphics workstation is obtained comparing the set of hardware benchmarks from the vendor. It is difficult to evaluate the true performance in this way because each vendor uses its own parameters; there are no standard low-level graphics benchmarks [10, 1]. We have written a set of high-level benchmarks using PHIGS, one of the best-known portable graphics systems.

In order to obtain representative measurements, we analyzed those parts of the graphics software which we frequently use in our applications. In Section 4.2 of this paper, we present the main concepts used by PHIGS. Section 4.3 describes the criteria we will evaluate, according to the PHIGS concepts and adapted to our applications. In Section 4.4, we present the results and analysis of our evaluations. Section 4.5 follows with comments about the PHIGS implementation; Section 4.6 with conclusions of the results and comments about the portability of PHIGS. Finally, we present future work in Section 4.7.

4.2 Main PHIGS concepts

PHIGS provides a functional interface between an application program and a configuration of graphical input and output hardware devices. The storage and manipulation of the data is organized in a centralized hierarchical data structure, where the fundamental entity of data is a structure element called *structure*. The creation and manipulation of the data structure is independent of its display. First, each object that is displayed must be represented in a structure or a set of structures. These structures are sent to an output workstation (normally represented by a window in a windowing system). Finally, by traversing the structures associated with the object, the output is displayed on the workstation.

The data stored during the creation of a *structure* includes specifications of graphics primitives, among which are: attribute selections, modeling transformations, view selections, clipping information, and invocation of other structures. When the post and update functions are called, the *structures* associated to the involved workstation are traversed; that is, each *structure* element is interpreted.

The geometrical information (coordinates) stored within the *structures* are processed through both workstation independent and dependent stages. The workstation-independent stage of structure traversal performs a mapping from modeling coordinates to a world coordinates system. The workstation-dependent stage performs mapping between four coordinate systems, namely:

- Word Coordinates (WC), used to define a uniform coordinate system for all abstract workstations,
- View Reference Coordinates (VRC), used to define a view,
- Normalized Projection Coordinates (NPC) used to facilitate assemblies of differents views, and
- Device Coordinates (DC), representing the display space of each workstation system.

4.3 Definition of the evaluation

In order to analyze our simulation results, we need to display 3D-surfaces, in general approximated via vectors, triangles, squares, and polygons.

The surfaces are obtained cutting a 3D data volume, and the polygons on the surfaces depend on the internal data structure, so the PHIGS routines for clip surfaces are inappropriate for this application. The surfaces must therefore be changed whenever the cutting planes are changed, and we must create a PHIGS structure and display it for almost every frame; this means the evaluation must consider the time to create a structure as well as the display time.

The program which evaluates the performance reads the input data necessary to define the type and size of the surface we want to evaluate. Each surface has only one type of polygon, i.e., square or triangle, and each figure has the same size. In addition, the program has the possibility to evaluate the computation time for vectors of different sizes. To display squares and triangles, the **Fill Area 3** output primitive is used and not **Set of Fill Area 3** because the last runs slower in the available implementation of PHIGS. To display vectors, the **Polyline 3** output primitive is used.

The performance time is computed in two parts:

- *Create_Structure* time, corresponding to the time between the **Open Structure** and **Close Structure** functions, and
- *Post_Redraw* time, corresponding to the time for **Post Structure** and **Redraw all Structures** functions.

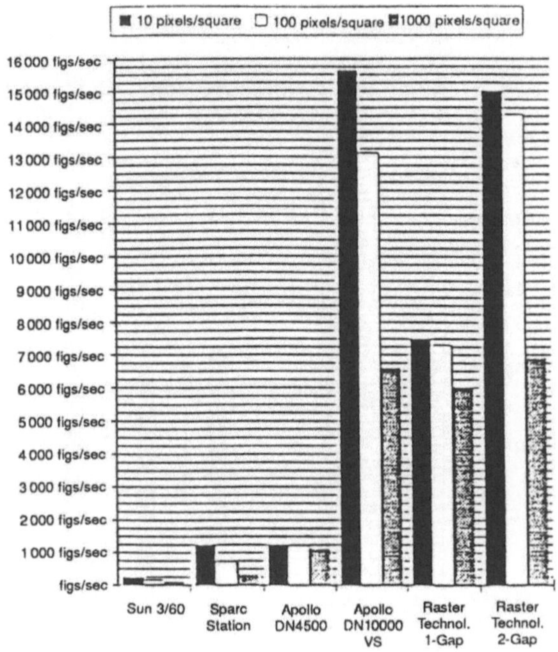

FIGURE 4.1. *Post_Redraw* speed for squares of different sizes

An *Overall* time is also presented, as described later in this paper.

In Addition, we were interested in comparing the different shading methods, namely *Flat shading*, where one constant color is used, and *Gouraud shading* , which applies the lighting calculation at each vertex of the fill area and then interpolates the resulting colors across the primitive. The last method is only possible if PHIGS+ is available.

4.4 Presentation of the results

Before we show the graphics benchmark performance, we describe some characteristics of the machines on which the performance of PHIGS has been evaluated. The machines are:

1. Sun 3/60 with 16Mb., Sun OS 4.0 and SunPHIGS 1.0.
2. Sparc Station with 16Mb., Sun OS 4.0 and SunPHIGS 1.0.
3. Sun 3/260 with 16Mb, & Raster Technologies GX4000 (1 and 2 Gaps), Sun OS 3.5 and PHIGS+ 1.0.
4. Alliant FX80 & Raster Technologies GX4000 (1 Gap). The Alliant contained 4 CE's and 3 IP's, but was used in detached mode (no parallel processing), Concentrix 5.0 and PHIGS 1.0.
5. Apollo DN10000VS with 64Mb., 1 processor, Unix Apollo 1.0. PHIGS 1.0
6. Apollo DN4500 with 16Mb., Unix Apollo 1.0 and PHIGS 1.0.

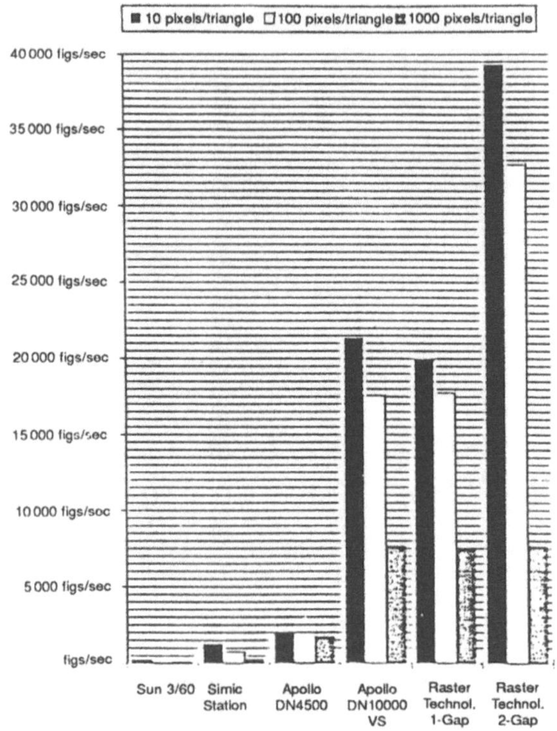

FIGURE 4.2. *Post_Redraw* speed for triangles of different sizes

The graphics benchmark performance is presented using the figures per second (figs/sec) unit. The first set of charts, figure 4.1, figure 4.2, and figure 4.3, show the *Post_Redraw* speed for squares, triangles and vectors of different sizes. In each case, we have taken approximately 10, 100, and 1000 pixels per figure. Both Alliant FX80 & Raster technologies 1 Gap, and Sun 3/260 & Raster Technolo have a similar *Post_Redraw* speed, so only one is presented. Those charts show a wide range of polygon and vector speeds, and indicate that the graphics capabilities of the supercomputing graphics workstation such as Apollo DN10000VS and Raster Technologies are many times faster than the other workstations.

The *Create_Structure* speed does not depend of the figure size. So, in figure 4.4, there are three bars associated with each machine; the first one represents the number of squares per second, the second one the number of triangles per second and the third one the number of vectors per second. The high performance shown above is lost here. The time spent in creating a structure is very large compared with the *Post_Redraw* time.

The *Overall* time is computed using results obtained for *Create_Structure* time and *Post_Display* time: *Overall* time = *Creat_Structure* time + *Post_Display* ti to compute the speed (figures per second), this time is divided by the number of figures displayed in that time. In the same way as the first set of charts, figure 4.5, figure 4.6 and figure 4.7 present now the overall performance of PHIGS. In our context, those charts show the Apollo DN10000VS performs best when displaying display squares and Alliant FX80 & Raster Technologies, triangles and vectors. It can also be seen that the addition of a

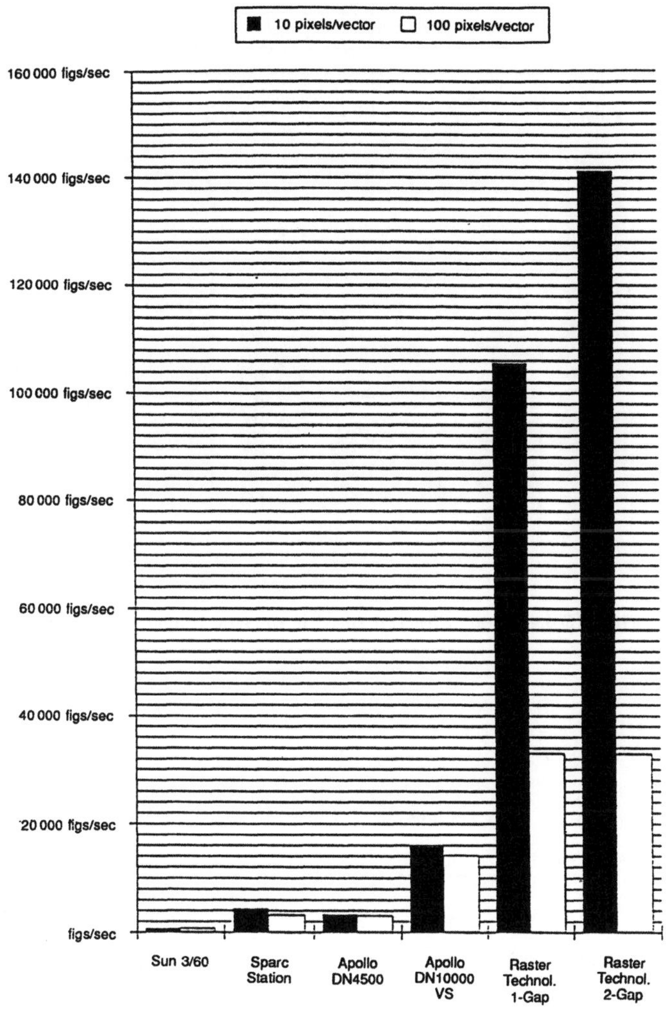

FIGURE 4.3. *Post_Redraw* speed for vectors of different sizes

second graphics processor in the Raster Technologies system brings little speed up, as most of the time is taken by the (sequential) *Create_Structure* code.

Finally, we present some measurements in order to compare the two shading methods; the *Post_Redraw* speed for the flat and Gouraud shading methods is presented. This comparison was only possible using the Raster Technologies & Sun 3/260 because PHIGS+ was available only on this platform. The *Create_Structure* speed is very similar for the equivalent output primitives, one using flat and the another using Gouraud

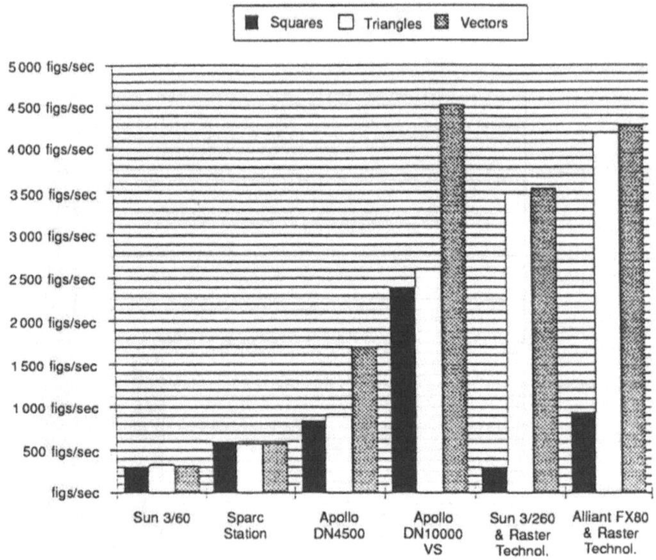

FIGURE 4.4. *Create_Structure* speed in squares, triangles and vectors.

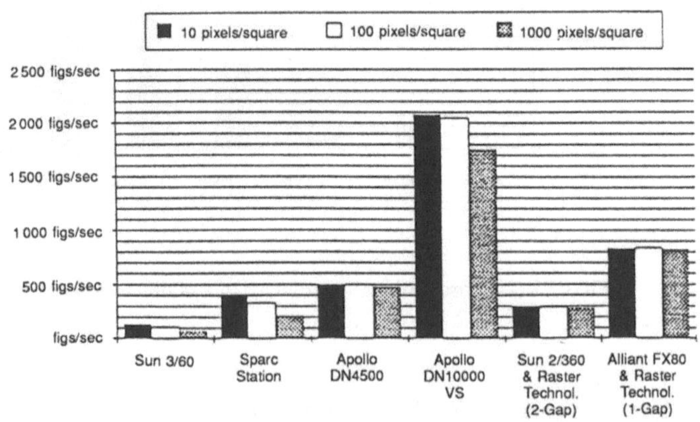

FIGURE 4.5. *Overall* speed in squares/second.

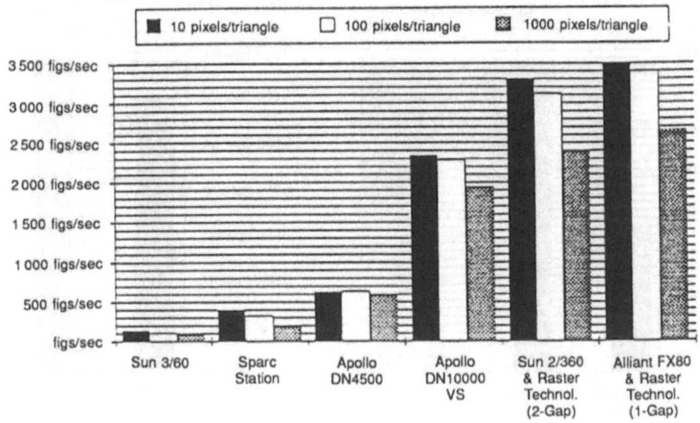

FIGURE 4.6. *Overall* speed in triangles/second

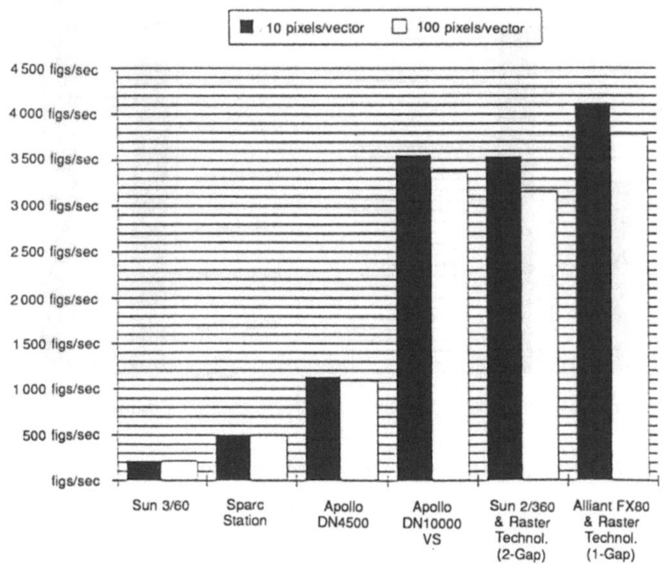

FIGURE 4.7. *Overall* speed in vectors/second.

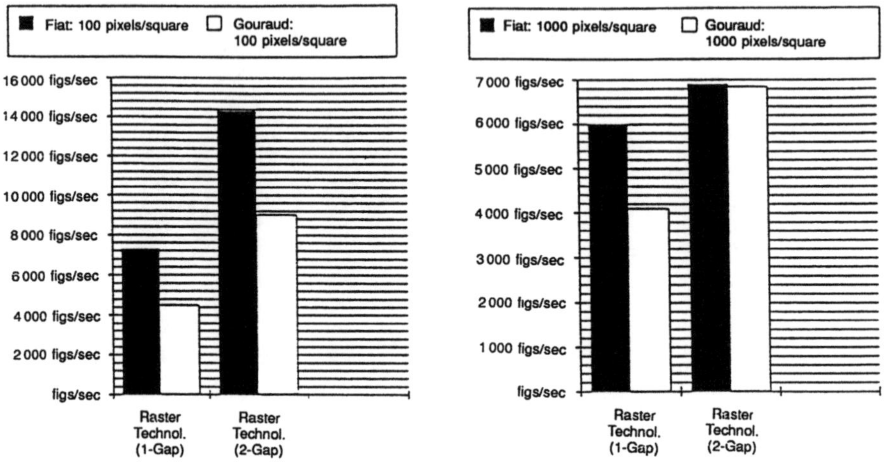

FIGURE 4.8. *Post_Redraw* speeds comparing the flat and Gouraud shading method in squares/second.

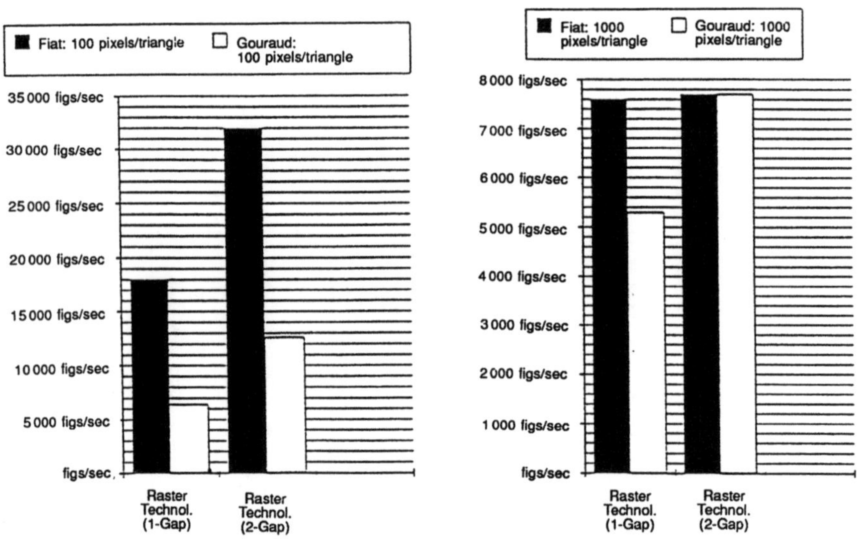

FIGURE 4.9. *Post_Redraw* speeds comparing the flat and Gouraud shading method in triangles/second.

shading method. For this reason, those measurements are not presented again. But the *Post_Display* speed associated to the output primitives using flat shading considerably faster than the output primitives using Gouraud shading. The next four charts show those measurements for squares, triangles, 100 pixels/figure and 1000 pixels/figure, respectively, in each case. Measureme 10 pixels/figure were not included because of its similarity with the 100 pixel

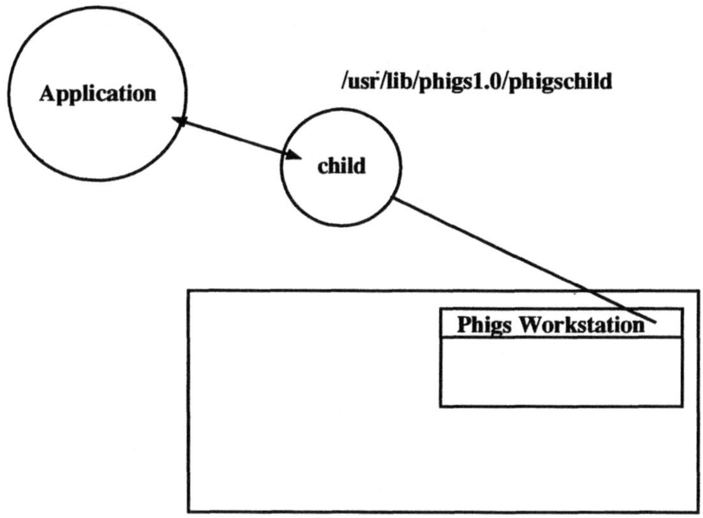

FIGURE 4.10. SunPHIGS implementation

4.5 Comments about the PHIGS implementations

In order to understand why SunPHIGS [13] is slower than it could be, in figure 4.10, a simple diagram of its implementation is showed. The application creates one child process for each opened workstation. In the available configurations, this schema does not bring advantages in the graphics performance, because on one hand, we must always wait until the display finishes, and on the other hand, the interchange of data between the processes slows the performance.

The PHIGS implementations on supercomputing graphics workstations take advantage of the powerful local capabilities. Supercomputing graphics workstations offer substantial computer-high-performance graphics. In the cases of the Apollo DN10000VS and the Raster Technologies workstations, the high-performance graphics performance is corroborated through the *Post_Redraw* time, shown in figure 4.1, figure 4.2, and figure 4.3. In figure 4.11, the graphics system configuration of the Raster Technologies[11] is shown in order to understand why creating a structure is very large in comparison with the *Post_Redraw* time in the *Create_Structure* process, the CPU creates the structures and puts them in the Display List Memory via VME bus. In the *Post_Redraw* process, the Display List Memory uses a high performance memory management processor with DMA Controller for the high speed list execution with minimal CPU intervention.

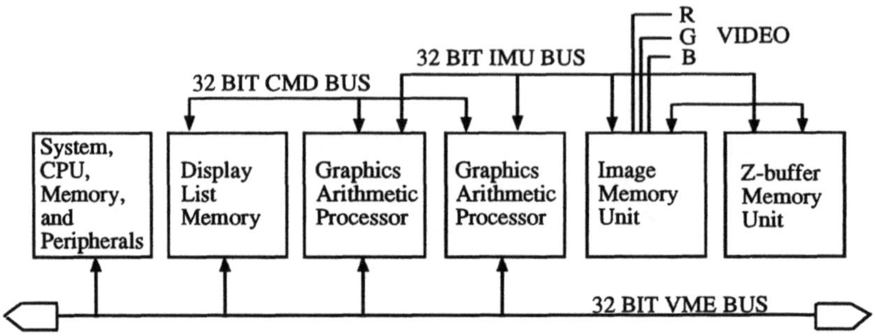

FIGURE 4.11. Graphics System Configuration of Raster Technologies

4.6 Conclusions and comments

Our original idea was to use PHIGS as the basis for our applications, mainly because of its portability. Additionally, we were sure that its performance on supercomputing graphics workstation would be good enough for our purposes. Through the measurements, the main problem is that the time spent to create a structure is very large, even on the mini-supercomputers. This suggests that in our type of applications it is not yet reasonable to use graphics software like PHIGS, because of the slow speed in creating large structures. Other types of applications, namely those which do not need to create structures very often, would be appropriate for implementation using PHIGS.

PHIGS provides the necessary functions to modify structures, to change a view (with parallel or perspective projections), and to apply transformations with reasonable speed. Still, the current versions of PHIGS are not bug-free. A frustrating problem with many PHIGS implementations at the present time involves the **Fill Area** function when used in 3D projections with polygons with more than three corners. The problem arises if the corners of the polygon do not lie in the same plane, and results in crashing the application, and sometimes also the workstation.

Another problem is PHIGS portability, because of its implementation on different systems. We have begun developing our performance program in the C PHIGS+ version, which runs on the Raster Technologies & Sun 3/260 [11]. The window system used here is NeWS [12]. The normal output workstation is a window of the window system. In this implementation of PHIGS+, the initialization values of the workstation, such as size, position, and category are defined through the **Escape** function, defined to manipulate non-portable characteristics. The constants defined for the identification of the input devices, for example the menu items, begin with the number zero.

The second step was to port the program onto a PHIGS version running on a Sun 3/60 with Suntools. The first problem was that SunPhigs uses the **Escape** function for another kind of implementation-dependent characteristics, and defines another PHIGS function to specify the original workstation attributes. This function is called **Workstation type set** and is used like a PHIGS function. The constants defined for identification of the input devices and also the menu item identification begin with the number one.

In the third step, we used the Apollo version of PHIGS. This version does not define a 3D point using the C language structures, where each point are implement through a x, y, and z fields. It is necessary to call the **Fill area 3** function using 3 arrays as parameters, one for each coordinate. This implementation looks like a Fortran version adapted to be used from C. Additionally, there are other functions to define the workstation characteristics, dependent on its window system.

Finally, we have tried to port our program to a Silicon Graphics, but only Figaro was available. It is similar to the Apollo version of PHIGS in the way the parameters are passed to the output primitives, but with new differences in the functions to specify a view. At the time of the writing, the C PHIGS version had not arrived.

4.7 Future work

The idea is to continue looking for graphics software and a graphics supercomputer which is able to satisfy our requirements. For example, it would be interesting to make some benchmarks to compare the performance of Doré graphics software, running on different workstations, with the PHIGS performance. For our 2D applications, we have made some benchmarks on 2D graphics software, but this work is still in progress.

At the same time, our group is developing a 3D graphics software using C++ with InterViews [8, 7, 9]. Soon we will be also able to obtain performance data for this software, which was written to match our requirements exactly. If its performance is much better than the performance of the portable graphics software, we will port our software to a graphics supercomputer. The software under development has been designed to be easily portable at the output primitive level.

Acknowledgements:

Thanks are due to Alliant Computer Systems, Apollo Computer, Ardent Computer, Sun Microsystems, Raster Technologies and Silicon Graphics for the provision of access to their hardware and for the assistance of their staff.

4.8 References

[1] K Anderson. Engineering Workstations: Technical Guide. *Computer Graphics World*, April 1989.

[2] ANSI. *Computer Graphics-Programmer's Hierarchical Interactive Graphics System (PHIGS)-Functional Description*, May 1987.

[3] ANSI. *Computer Graphics-Programmer's Hierarchical Interactive Graphics System (PHIGS)-Language Bindings*, March 1987.

[4] Ardent Computer Corporation. *Dorè Porting and Implementation Manual*, March 1989.

[5] Ardent Computer Corporation. *Dorè Programmer's Guide*, March 1989.

[6] J Burgler, P Conti, G Heiser, S Paschedag, A Aemmer, and W Fichtner. Multidimensional Semiconductor Device Simulation Algorithms, Implementation, and Results, 1989.

[7] M Linton. *Programming User Interfaces in C++ with Interviews*, 1989.

[8] M Linton and P Calder. *Interviews Reference Manual*, May 1989.

[9] M Linton, P Calder, and J Vlissides. *Interviews: A C++ Graphical Interface Toolkit*, 1989.

[10] G Marchant, M Stephenson, and T Crowfoot. A set of Benchmarks for Evaluating Engineering Workstations. *IEEE*, May 1989.

[11] Raster Technologies. *GX4000 User's Manual.*

[12] Raster Technologies. *GX4000 User's Manual.*

[13] Sun Microsystems. *SunPhigs Reference Manual*, 1988.

5 Visualization of Scientific Data for High Energy Physics: Basic Architecture and a Case Study

Carlo E. Vandoni

ABSTRACT
Visualization of scientific data although a fashionable word in the world of computer graphics, is not a new invention, but it is hundreds years old; examples of Visualization of Scientific Data are dating back since 1700. With the advent of computer graphics the Visualization of Scientific Data has now become a well understood and widely used technology, with hundreds of applications in the most different fields, ranging from media applications to real scientific ones.
In this paper, we discuss the design concepts of the Visualization of Scientific Data systems in particular in the specific field of High Energy Physics, at CERN. Then an example of a practical implementation is given.

> "... a computer display enables us to examine the structure of a man-made mathematical world simulated entirely within an electronic mechanism. I think of a computer display as a window on Alice's Wonderland in which a programmer can depict either objects that obey well-known natural laws or purely imaginary objects that follows laws he has written into his program.
>
> Through computer displays I have landed an airplane on the deck of a moving carrier, observed a nuclear particle hit a potential well, flown in a rocket at nearly the speed of light and watched a computer reveal its innermost workings". [1]

5.1 Introduction

Visualization of scientific data although a fashionable word in the world of computer graphics, is not new. On the contrary it is hundreds years old. An excellent book[19] gives an encyclopedic review of this technique. The author of this book presents many examples, some of them dating back since 1700. The author is quoting between 900 billion and 2 trillions images of statistical graphs printed every year in the world. With the advent of computer graphics the Visualization of Scientific Data has now become a well understood and widely used technology, with literally hundreds of applications in the most different fields, ranging from business graphics, to media applications and to real scientific ones. In the following, we discuss the usage of the Visualization of Scientific Data techniques and computer graphics in general and in particular in the specific field of High Energy Physics. Therefore, we first introduce CERN, one of the largest scientific research laboratories in the world. Then, the design concepts of Visualization of Scientific Data systems for High Energy Physics are discussed and an example of a practical implementation is given.

5.2 CERN and its computing facilities

CERN, the European Laboratory for Particle Physics, operated by the European Organization for Nuclear Research, one of the largest scientific research laboratories in the

[1] Ivan E. Sutherland, Computer Displays, Scientific American, Volume 222, Number 6, pages 56–81, June 1970

world, has as its primary function to provide European physicists with world-class particle physics research facilities which could not otherwise be obtained with the resources of the individual countries. This makes CERN the principal European centre for fundamental research in the structure of matter. The field of research is variously known as "particle physics" (since it is the study of the behaviour of the smallest particles of matter), "subnuclear physics" (since the particles involved are on a smaller scale than the atomic nucleus) or "high energy physics" (since high energies are needed to perform the research). It is "pure research" and is not concerned with the development of nuclear power or armaments. Most of the research work carried out in the Laboratory is strongly dependent upon the use of large particle accelerators, and computer systems of which more than 300, of various types, are presently installed. This multi-mainframe and multi-minicomputer complex is available daily to over 5,000 users, who are essentially physicists working on-site on various physics experiments.

The development of experimental techniques in high energy physics over the past 30 years has leaned heavily on the parallel development of electronic computers. In a typical experiment, hundreds of thousands, or even millions, of particle events are measured by using suitable detectors to observe the products of collisions between particles. After an event has been measured, its analysis often involves long and complex calculations. Computers entered the field of high energy physics very early on, and they are now so pervasive that there is almost no single building on the CERN site, where at least one computer is not present. In addition, the type of research carried on here has such unusual requirements for both computer hardware and computer software that it is often impossible to find suitable products on the market, and this has led to several locally-developed hardware devices and software packages.

Computers are used at CERN for many different purpose. Relevant to this paper is the usage in the context of the analysis of the experimental data. Modern High Energy Physics experiments produce an enormous data flow; a typical experiment writes several reels or cassettes of magnetic tape every hour. Although the situation is slowly changing with much larger computing power available on-line, today the full analysis of such volumes of data cannot realistically be performed on-line; data resulting from experiments written onto magnetic tape are analyzed on large mainframes either on site or at the collaborating Laboratories. The final product of an High Energy Physics experiment being in fact a publication of the results, the data resulting from analysis are very often presented in graphical form. Therefore, it is essential to make available to CERN users sophisticated and powerful graphics features; it is also highly desirable to link such a facility to desktop publishing systems. For this important application many different systems have been developed by CERN and by the collaborating researchers during the last twenty years.

5.3 Visualization of scientific data in the field of HEP

This kind of research has now virtually abandoned detecting techniques based on bubble chambers or similar devices, where physics phenomena were directly recorded on photographic films. Today, the data resulting from experiments are just streams of numbers recorded on literally thousands of magnetic tapes. An invaluable analysis technique is computer graphics, allowing complex data patterns to be visualized as two- or three-dimensional images. Graphics terminals and workstations are now integrated into all aspects of the Laboratory's activities, the principal activities being on-line monitoring, off-line analysis and event simulation. Today, the techniques for visualization of scientific data have become essential components of the chain of programs for HEP data analysis.

Needless to say, visualization of scientific data techniques are strongly dependent upon man-machine interaction. In these applications, we integrate interactively the power of a high performance workstation with the computing power and storage of the central computers.

Data analysis packages have been available to the HEP community for more than 25 years. Data presentation packages have been also available since a long time. In the past, the two functions were separated, as in any case the media available for presenting data in visual form were rather limited (for instance, for many years the only output medium normally available to produce a scatter plot was the line printer). However, it was recognized very early on that the provision of graphics facilities could be an important factor in helping the physicist in the analysis of experimental data and in the production of the results in graphical form. Nowadays, the availability of modern workstations, offering an adequate computing power and high quality graphics capabilities at an affordable cost, have had a major impact in the field of data analysis and presentation packages.

5.4 Visualization of scientific data: the four basic building blocks

At the basis of all systems for visualization of scientific data we always find four main building blocks, providing facilities for:

- data organization and retrieval
- data analysis
- data presentation
- user interface handling to control the whole application

There is a strong tendency today to integrate under the same system, for the visualization of scientific data, these facilities which in past were completely separated. By integration, we do not mean production of monolithic packages. This would be orthogonal to portability and modular development. The best solution is today to produce an homogeneous set of interfaces to the different facilities.

As any system for visualization of scientific data is by its own nature interactive, a fully interactive language, possibly organized in the form of a portable UIMS, understandable to the casual and to the inexperienced user, to make the system an effective tool in the hand of the physicist is needed. The syntax of the language should be particularly simple, and the system itself error-forgiving in the sense that errors must be detected immediately upon the entry of the command, and never fatal. Error messages should be absolutely clear and use a terminology which is familiar to the user. A powerful on-line help facility ought to be planned from the outset. Awkwardness of operating systems and graphic packages, whenever possible, should be hidden behind an appropriate human-engineered interface. On the other hand, it is important to permit the access to standard system functions, like the local editor, at any time, in a completely transparent way, with no need to leave the package to access the system function. The user should be able to define procedures and to store them under some name. He should be able to call on such a predefined procedure at some later time by means of its associated name. This facility will permit the easy development of individual sub-systems. Editing and manipulation facilities are also required, possibly using the same software tools standard as the system where the package is operating.

The availability of data is of vital importance. It must be possible to make data available to the system whether they reside on the same computer system or are stored remotely. Facilities to access data on heterogeneous networks are also very important.

5.5 An example system for the visualization of scientific data: PAW

PAW, a large software package where man-machine interaction and graphics play a key role, largely developed at CERN, is essentially an interactive system which includes many different software tools, strongly oriented towards data analysis and data presentation. Some of these tools have been available in different forms and with different human interfaces for several years.

PAW is conceived as a general-purpose instrument to assist physicists in the analysis and presentation of their data. A number of special-purpose systems of this kind have been developed in the past. It is worth mentioning here the MIDAS system[9] implemented at ESO, GEP[2], IDA [8], PV-WAVE[14], POL[1], and PUNCH [12] by RAL. PAW provides interactive statistical or mathematical analysis working on objects familiar to physicists like histograms, event files (Ntuples), vectors etc., coupled with powerful and sophisticated interactive graphical presentation facilities, allowing the production of graphical output of publication quality. PAW is primarily intended to be the last link in the analysis chain of experimental data. Some parts of PAW are also useful for graphical representations of simple data provided by the user, for the study of mathematical functions, and it may also be used as a data presentation package in a data-acquisition environment. It has, therefore, access facilities to experimental data, to HBOOK[6] data, or to any FORTRAN data, in addition to those generated by PAW itself, in numerical or graphical form. Powerful facilities for mathematical manipulation or combination of data objects (i.e. for data analysis), are also provided. As concerns data presentation, a large set of graphic functions is provided, easing the production by the user of graphs of high quality. This graphical part of PAW is essentially based on the features provided by HPLOT[15], coupled with the facilities provided by the new HIGZ[3] package.

PAW integrates most of the functionality of older packages such as HBOOK, HPLOT, SIGMA, MINUIT and COMIS. Several years of experience and hundreds of thousands of hours spent by users with these packages proved that the approach was a sound one. While integrating these packages, we managed to add functionality, to provide a full integration of the features offered, to permit full interchangeability of data and objects as well as full access to data bases.

The system has to be considered completely open-ended, in the sense that the user is able not only to personalize his own version, for instance using the mechanism of the KUIP macros, but he is also able to create his own interactive language adapted to his particular application, by using the command definition facility which is the basis of KUIP.

In order to quantify in some way the complexity of the system we could just quote that the total number of lines of code of the seven modules composing PAW is well over 100,000. This figure does not include the underlying graphics package, the mathematical routines or the data structures management system.

A detailed description of PAW appears in [5]. It is important to underline here the fact that the modules composing PAW, although fully integrated together, can be used autonomously or integrated in other packages. A typical example is the use of KUIP, ZEBRA and HIGZ in GEANT, [4]. All the components of PAW fully documented in the manuals quoted in the references.

5.6 The four basic facilities in PAW

5.6.1 Data organization and retrivial

The main objects onto which PAW is capable of operating can be categorized as in the following.

- Files: It is of vital importance for a package of this kind to have access to experimental data, no matter where the data reside. Therefore, facilities to retrieve data and to access them on the various machines available to the user are obviously part of the package. Files are, by their nature, non-volatile, and reside normally on disk areas.

- Histograms: Data resulting from HEP experiments are often presented in the form of histograms. The facility for producing this particular kind of graph is so important in HEP that the original semantics of the word "histogram" has been altered. An "histogram" in High Energy Physics jargon is not only "A representation of a frequency distribution by means of rectangles whose widths represent class intervals and whose areas are proportional to the corresponding frequencies" [2], but it means the complex of numeric and non-numeric information, enabling the data presentation package to build-up a histogram in graphical form. Histograms are volatile items, but they can be stored and retrieved by appropriate commands. Needless to say, all the functionality available under the well-known packages HBOOK and HPLOT is now available under PAW. In particular, this includes facilities for the creation of one and two-dimensional histograms, operations between histograms, projections and comparisons of histograms, etc.

- N-tuples: Ntuples can be considered as large named two-dimensional arrays. The Ntuples can be seen from the physics point of view as event files. The user can access the ntuple as a whole or single columns, or even single components. Columns can be identified by a name or by an index. A rather complete set of operators is available to deal with ntuples - these include capabilities to apply "cuts" or selection criteria to the ntuple data, using a notation where arithmetic and boolean operators and mathematical functions can be freely used. A powerful facility exists ('mask'), enabling the user to select an ntuple subset which has particular characteristics, allowing in this way very fast access to the data subset.

- Vectors: Vectors are in fact one to three-dimensional arrays of volatile nature. They can be created during a session, operated upon using the full functionality of SIGMA, used to produce graphics output, and, if necessary, stored away on disk files for further usage.

- Pictures: These are collection of graphical primitives or macroprimitives. They can be used to produce off-line hard copies but, more important, can be manipulated by the graphic editor available under HIGZ.

5.6.2 Data analysis

The facilities offered by a number of widely used packages have been integrated into PAW: HBOOK, MINUIT, SIGMA, and COMIS.

[2]Webster's Third New International Dictionary of the English Language, Merriam Webster, 1981

HBOOK [7], and[6] provides the basic functionality to handle histograms, its main purpose being to define, fill and edit histograms, scatter plots and tables. The main application of HBOOK is to summarize basic data derived from experiments or from the subsequent analysis process. It can also be used to represent real functions of 1 or 2 variables. A number of minimization and parameterization tools is also available. Essentially for historical reasons, the basic output was originally formatted for the printer, but a graphics interface to the powerful PAW graphics options is provided.

MINUIT [11] is conceived as a tool to find the minimum value of a multi/parameter function and analyze the shape of the function around the minimum. The principal application is foreseen for statistical analysis, working on chisquare or log/likelihood functions, to compute the best/fit parameter values and uncertainties, including correlations between the parameters. Its functionality is now fully integrated in PAW, where it is used for histogram and vector fitting. A fully interactive version is also callable by PAW.

SIGMA (System for Interactive Graphical Mathematical Applications) [10, 18] provides the handling of mathematical expressions on vectors and arrays. SIGMA, which has to be considered as a system for interactive on-line numerical analysis problem-solving, has been designed essentially for mathematicians and theoretical physicists. It has been operational at CERN for several years on CDC CYBER computers, and its main features have been now incorporated into PAW. The basic data units on which the system is able to operate are scalars, one-dimensional arrays, and multi-dimensional rectangular arrays; SIGMA provides automatic handling of these arrays. The calculation operators of SIGMA closely resemble the operations of numerical mathematics. A subset of the CERN Library mathematical routines is accessible via SIGMA. SIGMA is also used in many places in the system in order to evaluate mathematical expressions occurring in many different PAW commands.

COMIS [20] is a FORTRAN interpreter, allowing the user to write and execute in interpretive mode FORTRAN subprograms. This facility is of great importance, as in this way the user has the possibility to write his own data analysis procedures, for instance his own selection criteria, minimization functions, etc..

5.6.3 Data presentation

In the past the preparation of graphs of publication quality has always been considered a tedious job and very often this was delegated to draftsmen. The graphics facilities offered by PAW, linked with the availablity of top quality hard copiers (like modern laser printer/plotters), permit even the inexperienced user to produce graphs ready for publication. The final goal of PAW is to produce publication-quality graphs, and it is therefore essential to make available the most sophisticated and powerful graphics features. It is also important to have these facilities linked to desktop publishing systems. A large spectrum of graphics options is available at the user's fingertips: histograms, scatter plots, contour plots, error bars, bar charts, column charts, pie charts, surface and "lego" plots, etc. In most cases, and certainly for the inexperienced or casual user, graphs can be produced by just providing a minimum of information, e.g. where the data are stored and what kind of graph is desired, and then the system produces automatically the graph, using default parameters and with no need for further programming effort. However, the more experienced user is seldom satisfied by graphs produced automatically, and therefore many facilities are provided, permitting the user to alter the default parameters related to the graph, enabling him to produce a fully personalized picture.

HPLOT is a FORTRAN-callable facility for producing HBOOK output on graphics devices. Its main design objective is to be able to produce drawings and slides of a quality suitable for talks and publications. The functionality of HPLOT has now been incorporated into PAW in a fully interactive mode.

HIGZ The use of graphics packages like GKS is becoming increasingly important for the provision of a standard interface between user programs and devices. The use of only one such package, GKS, provides portability of application programs between systems on which GKS is installed, and makes the application programs largely device-independent. These packages, however, have limitations. They do not provide the high level functions (axes, graphs, logarithmic scales, etc.) necessary for a data presentation system. There are always (sometimes minor) differences in the actual implementations on different computers. They do not foresee an acceptable way of recording large volumes of graphical information in compact form with a convenient access method for later manipulation.

The package HIGZ (High level Interface to Graphics and ZEBRA) [3] is an interface package between the user program and an underlying graphics package. It provides an interface to a standard data structure management system (ZEBRA) and through it a mechanism to store graphics data in a way which makes their organization and subsequent editing possible and easy. The picture data base is highly condensed and fully transportable. A picture editor is part of the package, allowing merging of pictures, editing of basic graphics primitives, operations onto HIGZ structures, etc. HIGZ is an interface package aiming at graphics applications of any nature, provided the level of functionality is similar. The package is basically a *thin layer between the user program and an underlying graphics package.* The level of HIGZ was deliberately chosen to be close to GKS and as basic as possible. This makes the interface to GKS a very simple one; and preserves full compatibility with the most important underlying graphics packages. HIGZ does not introduce new basic graphics features, and does not duplicate GKS functions.

The system is articulated into four main sets of functions:

- Basic graphics functions, interfacing to the underlying graphics package, with calling sequences identical to those of GKS.
- Higher-level macro-primitives, and the related control routines.
- Data structures management functions, interfacing to the data structures management system (ZEBRA).
- Picture editing functions.

The user is able to invoke any of these sets of functions, simultaneously or not. This is particularly useful during an interactive session, as the user is able to "replay" and edit pictures previously created, with no need to recall the application program, but just accessing the picture data base. On the other hand, some graphics macro-primitives are implemented, providing very frequently used functions such as histograms, full graphs, circles, bars and pie charts, axes, etc.. In addition, facilities to draw surfaces, contour plots, lego plots, etc. are also available. HIGZ is presently interfaced to several versions of GKS, to GL(Silicon Graphics, IBM RISC

6000), to GPR/GMR, to CORE/DI3000, and to X-Window. The underlying graphics package is chosen at compilation time. HIGZ provides both 2D and 3D graphics features. In addition, a PostScript driver and communication facilities (TELNETG) are provided. The TELNETG program permits the running of an application based on HIGZ on a remote computer (mainframe) and the performance of the graphic input/output on a local computer (workstation), combining the power of the CPU of the mainframe and the flexible graphics facilities of the workstation.

5.6.4 User interface handling

KUIP (Kit for a User Interface Package) [17] represents a new approach to the relatively old problem of general-purpose User Interface systems. The basis of KUIP is the so-called Command Definition File (CDF) which constitutes a very concise formal description of the command structure. An appropriate module, the KUIP Compiler, generates from this file a set of FORTRAN subroutines which are then compiled and linked with the interactive application.

The dialogue between the user and the system is handled either by typing command lines or by choosing alphanumeric or graphical menus. In order to avoid verbosity in typing command lines, often useless and annoying for the experienced user, command abbreviations are possible as long as they do not produce ambiguities. The user is able to switch from one dialogue mode to the other at any moment. Furthermore, menus do not need any special additional programming; they are automatically derived from the command structure as described in the CDF. Needless to say, it is possible to group commands in 'macro' files and invoke these sets of commands for repeated execution. Assignment and control statements may be used inside macros. KUIP keeps track of all command lines entered (independently of the dialogue mode), which are automatically recorded in a macro file, and may therefore be edited and re-executed as with normal macros. An Unix-like history mechanism is also available.

An on-line help facility is integral part of the system. This can be considered as a good step in the direction of the "system to be used without a manual". The importance of proper documentation should never be underestimated: it is vital to make available users' documentation since the release of the first prototype of the package (documentation which could possibly be produced automatically by the system itself). It is also very important to provide different levels of documentation: beginners' guide, users' guide, reference manual, each one having a different target and different users. One of the major problems normally encountered in this area is how to keep the documentation up-to-date. KUIP solved this problem by introducing a feature in the system which permits the automatic generation of the documentation (already in text processing format). As an additional benefit, as the documentation is an integral part of the language description system, the updating of the documentation is automatic. To our knowledge these two features (automatic production of documentation and automatic update) are features unique to PAW.

5.7 An important aspect of the software development: the portability

During the last twenty years, CERN has played a leading role as the focus for development of packages and software libraries to solve problems related to High Energy Physics. The results of the integration of resources from many different Laboratories can be expressed in several million lines of code written at CERN during this period of time, used at CERN and distributed to collaborating Laboratories. Nowadays, this role of software developer and distributor is considered very important by the entire High Energy Physics community. In the process of developing software to be used in many different places, several key goals have to be met and a number of criteria and rules are to be observed. The most important one is felt to be the portability. Computers used at CERN and in the collaborating Laboratories are all but homogeneous. In practice, it is neither possible nor desirable to impose standards as concerns the hardware to be used at CERN and in all collaborating Laboratories. Many different manufacturers and product lines are involved; just to quote some of the most important ones involved, we could mention IBM, DEC, Apollo and Cray.

Punched cards were finally phased out on the CERN site only at the end of 1986, when all CERN users were converted to terminal access to computers in a time-sharing environment. We are just starting the next phase, consisting of the elimination of dumb terminals and their replacement with powerful workstations, backed up by a few very large machines, still needed for the provision of large computing power for batch processing and of large backing storage. This will take us to the next millennium. In this kind of environment, an analysis and/or data presentation package must be able to run equally well in *batch processing* on the *mainframe*, *interactively*, via dumb terminals in *time-sharing* on the mainframe, and *interactively* on the *workstations*. Therefore portability is the password for the successful implementor of software packages. There are several aspects of portability that should be taken into account: portability of the source code, support of the code, interface with the operating system, interface with the local data structure and storage, etc. This general need led CERN quite a long time ago to develop also portable software tools, e.g.: source code management systems, data structures management systems, mathematical and utility libraries, etc., which are either completely portable or machine-independent. Of course, one natural issue is the adoption of Standards. When a large number of computers of different size and brand are used on the same site, it is important to adopt standards on those computers. Probably the only area where standards were very quickly adopted within the High Energy Physics community was that of a common programming language. The main reason was just historical: at the time when computing entered High Energy Physics applications no viable programming language other than FORTRAN was available [13]. Today, FORTRAN is virtually the only programming language used by physicists at CERN, and it is in fact used by them as in the past the slide rule was used by the engineer. As for FORTRAN, similarly development on standard graphic packages started very early at CERN (in 1964). With the adoption of a true international standard, GKS, CERN decided also to conform to this standard.

The package has been implemented in a *totally portable* way. This was achieved by making a large use of FORTRAN 77, of a general interface (HIGZ) to standard graphics packages and of a standard data structures management, ZEBRA. The package is also fully transparent to the operating system, permitting a direct interface to the systems tools.

5.7.1 Distributed PAW

Another unique characteristic of PAW is its ability to run in a distributed fashion. In practice, the user sitting at his workstation is capable of accessing files on remote computers, even if the data representation is different. This is achieved thanks to the object-oriented system ZEBRA [16] and a Remote Procedure Call facility built on top of TCP/IP or equivalent transport protocol. In an homogeneous environment, a PAW user can also access data coming from a different process via a memory mapping technique (i.e., global sections on VAX/VMS). The facilities have proved to be very useful in data acquisition systems and will become vital in a world of distributed computing.

5.7.2 Usage of PAW

PAW has been installed today in over 200 laboratories in the world. It is almost impossible to say how many users exist today but our estimation is that their number is well above 3000. At CERN a recent measurement has shown over 1000 users and over 50000 sessions per month.

PAW is fully implemented on the following systems:

- Apollo
- DEC Station
- IBM, VM/CMS and MVS/TSO
- IBM RISC 6000
- Sun
- Silicon Graphics
- VAX/VMS
- and on any UNIX platform

Partial implementations also exist for: CDC, CONVEX, Cray, NORD, and UNIVAC.

5.8 Conclusions

We have discussed in some detail the architecture of systems for Visualization of Scientific Data in particular in the area of the High Energy Physics. The present situation at CERN was presented, and the development work done at CERN in connection with these systems was outlined.

Acknowledgements:

The author wishes to thank Rene Brun, who has the direct responsibility for the development of many application packages at CERN for his very important comments and contributions to the writing of this paper.

5.9 References

[1] N Armenise, G Zito, and A Silvestri. POL: An Interactive System to Analyze Large Data Sets. *Computer Physics Communications*, 16:147–157, 1979.

[2] E Bassler. *GEP User Manual.* DESY, 1985.

[3] R Bock, R Brun, O Couet, R Nierhaus, N Cremel, C E Vandoni, and P Zanarini. *HIGZ Users Guide.* CERN Program Library.

[4] R Brun, F Bruyant, M Maire, A C MacPherson, and P Zanarini. *GEANT3.* CERN Program Library.

[5] R Brun, O Couet, N Cremel, C E Vandoni, and P Zanarini. *PAW-Physics Analysis Workstation.* CERN Program Library.

[6] R Brun and D Lienart. *HBOOK Users Guide.* CERN Program Library.

[7] R Brun and P Palazzi. Graphical Presentation for Data Analysis in Particle Physics Experiments: The HBOOK/HPLOT Package. In Carlo E Vandoni, editor, *Proceedings Eurographics '80*, pages 93–104, Amsterdam, 1980. Eurographics, North-Holland.

[8] T Burnett. IDA Interactive Data Analysis. Technical Report SLAC MAC-III memo 1/83-6, SLAC, 1983.

[9] Image Processing Group European Southern Observatory. *MIDAS Users Guide.* ESO, Garching bei Munchen, January 1988.

[10] R Hagedorn, J Reinfelds, C E Vandoni, and L Van Hove. *SIGMA, A New Language for Interactive Array-orientated Computing.* CERN Program Library, 1973/78.

[11] F James and M Roos. *MINUIT, Function Minimization and Error Analysis.* CERN Program Library.

[12] R T Lawrence, W C A Pulford, and M A Sturdy. *PUNCH User Guide.* Rutherford Appleton Laboratory, Chilton, Oxon, OX11 0QZ, UK, December 1988.

[13] M Metcalf. Aspects of FORTRAN in large-scale programming. Technical Report CERN/DD/82-18, CERN, November 1982.

[14] Precision Visuals Inc, USA. *PV-Wave Precision Visuals Workstation Analysis and Visualisation Environment-Introduction*, June 1988.

[15] N Cremel R Brun and O Couet. *HPLOT Users Guide.* CERN Program Library.

[16] R.Brun and J.Zoll. *ZEBRA Users Guide.* CERN Program Library.

[17] R.Brun and P.Zanarini. *KUIP Users Guide.* CERN Program Library.

[18] J Reinfelds and C E Vandoni. Sigma 76. In B Gilchrist, editor, *Proceedings Information Processing '77*, pages 963–978, Amsterdam, 1977. IFIP, North-Holland.

[19] E R Tufte. *The Visual Display of Quantitative Information.* Graphic Press, 1983.

[20] V.Berezhnoi, R.Brun, S.Nikitin, Y.Petrovykh, and V.Sikolenko. *COMIS Users Guide.* CERN Program Library.

6 The IRIDIUM Project: Post-Processing and Distributed Graphics

D. Beaucourt, P. Hemmerich

6.1 Introduction

This paper presents one of the latest project in scientific visualization undertaken by the Direction des Etudes et Recherches, the research and development department of Electricite de France, the French national electricity company. The aims of this project, called Iridium, are to develop a new post-processing tool for visualization in fluid dynamics and more generally to investigate how to take advantage of cooperative architectures including supercomputers and graphics workstations to achieve more powerful systems in the scientific area. The development of the post-processor itself will be the first experiment in which we will be involved in distributing software and we hope that it will be helpful for further applications. We maintain that distributing a scientific application on a supercomputer and one or more workstation connected together through a high speed network ideally responds to the requirements of scientific visualization.

6.2 What is required in visualization of fluid dynamics

A fluid mechanic code such as N3S developed by the EDF, computes data such as fluid velocity, pressure, turbulent kinetic energy, etc. These data are fields, either scalar fields or vector fields. The values of the fields are defined at each point of the physical domain - it is supposed to be a continuum. Yet the computational code computes the values of the fields only at some points of the domain, the nodes of the finite element mesh used by N3S. We propose to visualise these fieds to enable the user to readily understand the fluid flow. The problem in this kind of visualization is to find graphic representations well suited to continuous fields especially three-dimensional fields. But before asking for visualization of abstract data, the user simply wants to see the material object he is studying. It could be a pump or the wing of a plane. Let us first address the visualization of geometrical objects defined by meshes.

6.2.1 Visualization of meshes

The meshes represent either a concrete object or more abstract surfaces such as isovalue surfaces extracted from 3D-fields. The meshes are finite element meshes composed of tetrahedrons or bricks. The problem consists in rendering the meshes to represent surfaces in a natural way. The following visualization options are required: - perspective view - removal of hidden lines or surfaces - lighting and shading (flat or Gouraud shading) - wire frame representation

We believe however that the user can have a good understanding of a complex or unknown geometry only if he also has the ability to rapidly manipulate it, to systematically explore how it looks from different angles. A high interactivity speed is more useful than the realism of the images. A clipping plane interactively moved by the user is very useful in analysing a complex geometric system composed of many objects, some hidden by others. The possibility of selecting an object and of making it visible or not is also very useful in that case.

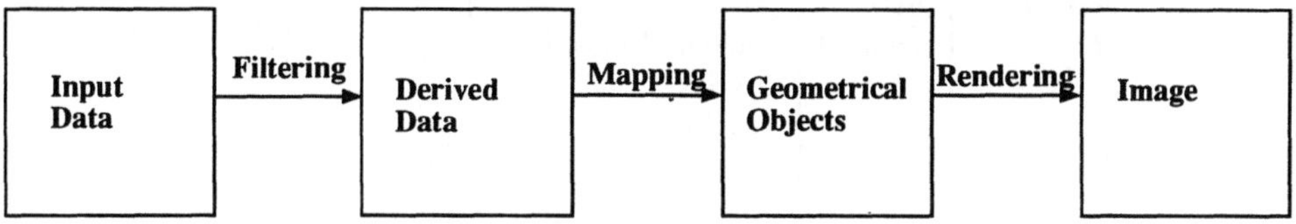

FIGURE 6.1. The visualization process

6.2.2 Visualization of fields

The vizualisation process

It is necessary to keep in mind the different transformations that convert numerical data into a displayable image. We briefly present these transformations. For futher details see [1] and [2].

We can define four major steps in the visualization pipe-line (see figure 6.1):

Computation of derived data: filtering In this step one finds numerical calculations such as integral, gradient, and interpolation. In some cases , these calculations can be time consuming. The result of this step is a set of new variables more condensed or more relevant than the input data.

Creating geometric objects: Mapping The numerical data are then converted into a geometric object. A geometric object is composed of geometric primitives: points, lines, surfaces or volumetric cells (voxel). The following section shows which transformations are useful in fluid dynamics.

Rendering The rendering step transforms geometric objects into an image. The user can assign visual properties to each primitive in order to obtain a more or less realistic image. As already mentionned sophisticated rendering algorithms are not useful because a visual realism is nor required.

Playback The rendering process produces images that are not necessarly displayed. If we want to understand the evolution in time of a phenomenon a snap-shot image is insufficient. We need animated images. An animation can be obtained by playing the computed images at different speeds. As the images have been computed in advance, it is possible to sequencing them fast enough to get a smooth animation.

The problem of volumetric visualization

To visualize a 3D continuum field is as difficult as to see inside a real concrete object made of some material. Unless the material is transparent, we can see nothing but the object's outline. This is not the entire object this is only a small part of it, its boundary surface. If we would like to see inside the object we need to break it up, to cut it and to examine a cross-section. However here again we would be dealing only with a surface not with a volume. Were the material transparent we could see the object's entire volume. A radiography in fact works in this way. In Iridium we don't use the transparency to render volumes. We don't really visualize 3D-fields, we transform these fields into simpler ones

Dimension of the domain	Geometrical primitives used			
	Points (0D)	Lines (1D)	Surfaces (2D)	Volumes (3D)
1D	coloured points	- coloured lines - y=f(x) - profile		
2D	coloured points	coloured contour lines	- coloured maps - z=f(x,y)	
3D	coloured		- iso-surfaces	voxel

TABLE 6.1.

by reducing the volumetric domain to non volumetric domains such as surfaces, lines or points. Only fields defined on these subdomains are then visualised. Subdomains include the following:

- the boundary surface
- a cutting plane
- a 2nd order surface (sphere or cylinder are very useful in axisymmetric problems)
- lines or points
- any other surface or line built from fields (isovalue surface or particle trace)

It is essential to enable the user to displace a surface as continuously and as quickly as he may wish, so that he may reconstruct a mental picture of the whole volume. It is equally important for users dealing with very twisted surfaces such as turbine blades: to be able to transform these convoluted surfaces into flat ones where all is visible.

Visualization of scalar fields

Table 6.1 gives an overview of the different possible representations.

It is possible to make a distinction between the representations made in the domain itself and others made in more abstract spaces. In the first case the principle used is very classical It consists in coloring the domain of a field, or some part of it, with colors depending of its values through a transfer function defined by a palette. Generally the colored domain is a surface that the user has selected and which he can move through the entire volume. It can also be the volumetric domain itself. In this case the transfer function is assumed to be discrete and the system creates colored surfaces, the isovalue surfaces. The user can move these surfaces by changing the corresponding values in the palette. Again the interactivity speed is a critical aspect of the system because we would like this transformation be continuous. In the second case, we use a coordinate system to make more classical representations like a y=f(x) curve or a z=f(x,y) surface. We believe that it is advisable to separate the two kinds of representation in different windows of the screen. Therefore a window system like X11 can be very useful here. The exchange of data between windows must be possible. For instance we would like to capture data in the 3D-domain and represent them on a curve, plotted in another window. This technique makes it possible to develop an especially open system.

Dimension of the domain	Geometrical primitives used			
	Points (0D)	Lines (1D)	Surfaces (2D)	Volumes (3D)
1D	location of a single particle	- coloured lines - y=f(x) - profile		
2D	location of a single particle	-arrows -trajectory of a single particle -displacement of particles starting from a line		
3D	location of a single particle	- arrows - trajectory	- displacement of particles starting from a surface ribbon (trajectory of 2 or more particles)	voxel

TABLE 6.2.

Visualization of vector fields

Table 6.2 shows the different representations possible. As the vectors represent the velocity, it is possible to define particle traces injected in the stream or their location at any given instant. The vectors can have been processed before visualization. Projections ,the magnitude of vectors or other quantities can be computed. Consequently it is interesting to use the same geometric object to represent a scalar field (the magnitude of vectors) and a vector field at the same time. A particle trace can thus be colored by the magnitude of the vector velocity or any other scalar field. Different techniques are required to reduce the number of displayed vectors. Only the vectors at the nodes of a given grid or only short vectors are drawn for example.

Yet for both fields, we must keep in mind that an image only represents fields at a given instant ant that what the user actually needs to see is the evolution of the phenomenon in time. The user therefore needs animated visualizations. We believe that a post-processor must offer the possibility of defining scripts of animation sequences producing a set of images one could at least record one at a time on a video tape for a later playback. Nonetheless it would be much practical if the animation could be visualized immediatly on the workstation itself. This would require a very compute intensive system. We affirm that a high performance graphics workstation connected to a super computer through a high speed channel is a possible solution.

6.3 The user interface

As we want the post-processor to be easy to use, the user interface must be carefully designed. A user interface has two aspects, a semantic one and a syntactic one. The semantic aspect defines the functions of the system i. e. what one can do with it. The syntactic one defines how a functionality can be achieved i.e. the sequences of physical actions the user should do to activate a given function. We have pointed out that the visualization functions of the system work correctly only if they are put under the interactive control of the user. The problem of the user interface is not the specification of its functionalities but rather the translation of a given function into a sequence of physical actions with the input devices. What we want to do is to make these actions so intuitive and natural that the user has the illusion of working directly on the geometric objects not on a mouse or a keyboard. There are very interesting tools in the X window environment such as pop-up menus, buttons, dialog-boxes, etc. However particular problems arise in the manipulation of 3D geometrical objects. The manipulation of a selected objet in the 3D-space is an example. The interactivity speed is again a crucial point because an immediate feedback after each command is necessary: this way the user can always see what he is doing anf if necessary immediately correct his actions.

6.4 The implementation

6.4.1 Cooperation supercomputer - workstation

Scientific visualization requires not only a minimum of drawing speed but also a minimum of computing speed to get effective interaction. Furthermore, in this area we find large volumes of data. The meshes processed by N3S can reach 50000 nodes. At each node up to 10 values can be computed corresponding to 100 time-steps. Consequently the power of a single workstation might be not sufficient. In addition we think that a rigourous separation between the computational code and the visualization tool is in some cases not recommended. For example, to see what happens in fluid dynamics when something is introduced or moved in a windstream, the system must recompute the data every time the geometry changes. A tighter coupling between calculation and visualization would in fact be more effective consequently a cooperation between a workstation and a supercomputer could provide a satisfactory solution.

6.4.2 Utilization of standards

To shorten the development time and to achieve portable applications we need development tools available on most computers. The choice of standards has been decided.

- PHIGS/PHIGSPLUS in graphics
- X-Window and the MOTIF toolkit as tools to develop the user interface
- NFS to share the discs and files
- different unix tools (RPC, Sockects, pipes,etc) to share the processing

6.4.3 Possible architectures

Figure 6.3 shows a variety of architectures with different distributions of the software between a Cray computer and a workstation. Each one has its advantages and its drawbacks.

Architecture 1 is probably the simplest. The application takes place entirely on the workstation. There is no distribution of software so it is easy to implement. As the files reside on the discs of the supercomputer it is possible either to transfer them for example by a ftp command or to use NFS to share the Cray discs. Two bottlenecks are possible: filtering and mapping. Both use the workstation cpu which can be slow to process compute intensive functions like extracting isovalue surfaces from a volumetric field. The rendering step must be achieved by a specific hardware to give good performance.

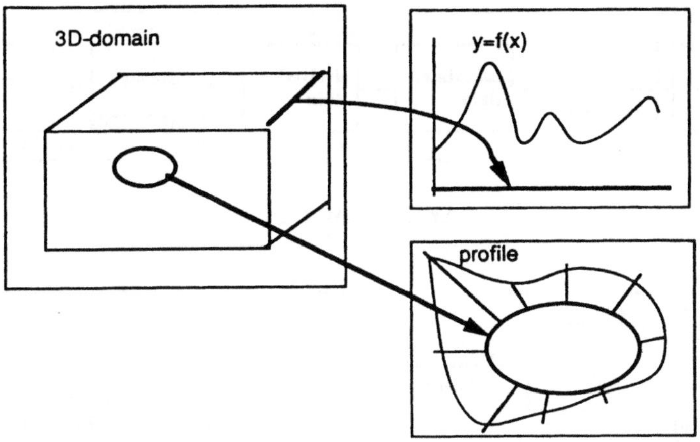

FIGURE 6.2. Different representations of the same data in different windows

Architecture 2 removes some compute intensive functions from the workstation to put them on the supercomputer.The communication between supercomputer and workstation can be based on RPCs or sockets. The network can be a bottleneck although the volume of data to transfer is relatively limited.

Architecture 3 is probably the most efficient. The workstation is only used to render geometrical objects. that have been computed as fast as possible on the supercomputer. Nevertheless the workstation can be a bottleneck if it has not been well designed: the cpu might not be able to transfer the data fast enough from the network to the graphics engine. The workstation must be well balanced : a powerful cpu a rapid visualization pipeline and a high speed bus between both.

Architecture 4 is based on the same principle as architecture 3 but instead of being used as a 3D visualization server the workstation is a simple X11-server. It can be regarded as a temporary solution as long as PEX is not available. It also has the advantage of a low cost as any workstation or X-terminal can be used. The rendering however might be poor since it would include no shading or lighting.

In architecture 5 almost everything is processed in the supercomputer. The workstation is only used as a frame buffer. The rendering step and the image transfer are possible bottlenecks. Can a supercomputer be faster than the specific graphics engine of any workstation? We have no answer at this time. The image transfer seems to be the most critical step. One has to transfer 1000x1000x12 bits for an image. An ethernet link is not suitable to frequent transfers of 12 megabits. This architecture could be a solution for animation problems. The supercomputer computes every image of the animation and stores them on disc. Once all images have been computed it is possible to make an animation by transferring them to a frame buffer at the rate of 25 images per second. A very high speed network is required: 12 megabits x 25 = 300 megabits/second which is feasible. In some cases it might be possible to achieve animations without computing the images in advance.

We intend to design the application without a priori.choosing one of the previous architectures The architecture will be choosen at the installation time not before. If necessary we will develop application programmer interfaces on the top of the RPC or sockets to make the network transparent.

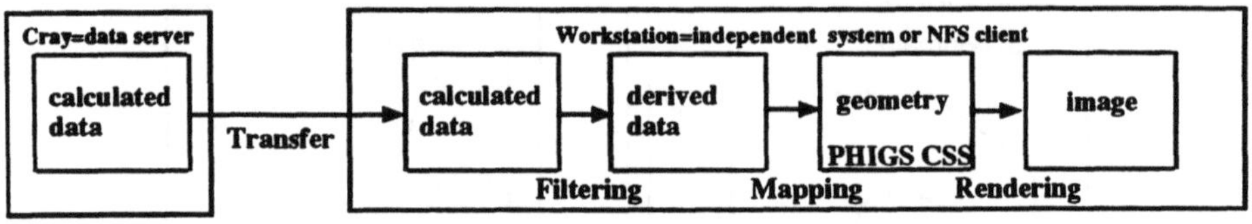

Architecture 1

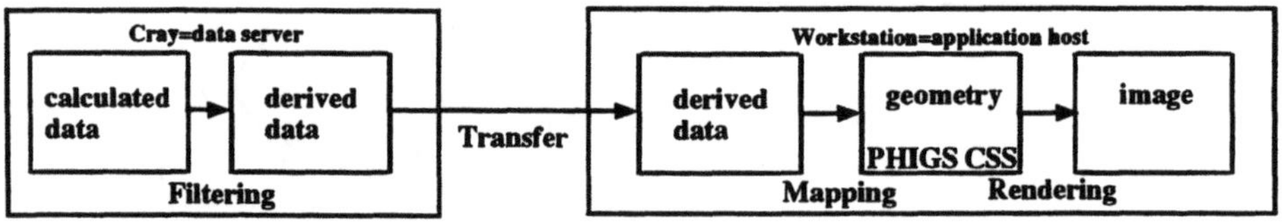

Architecture 2

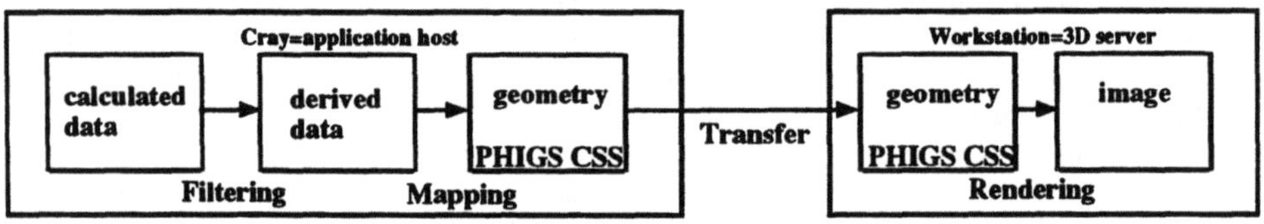

Architecture 3

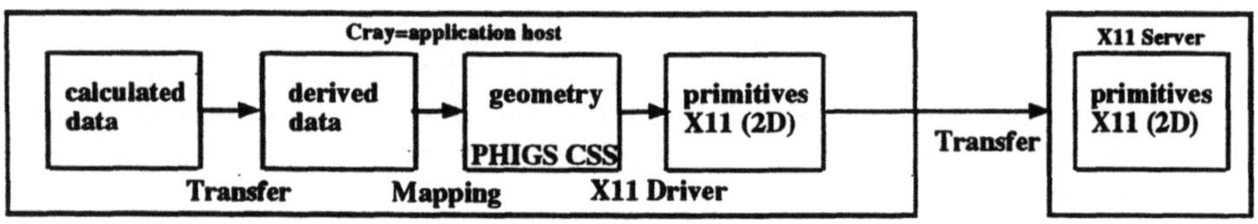

Architecture 4

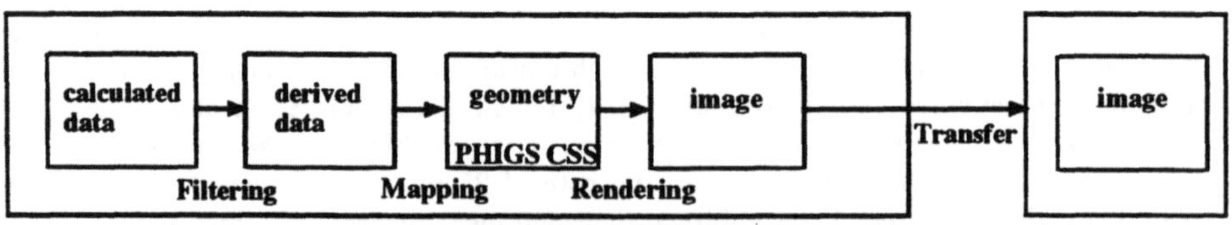

Architecture 5

FIGURE 6.3. Possible architectures

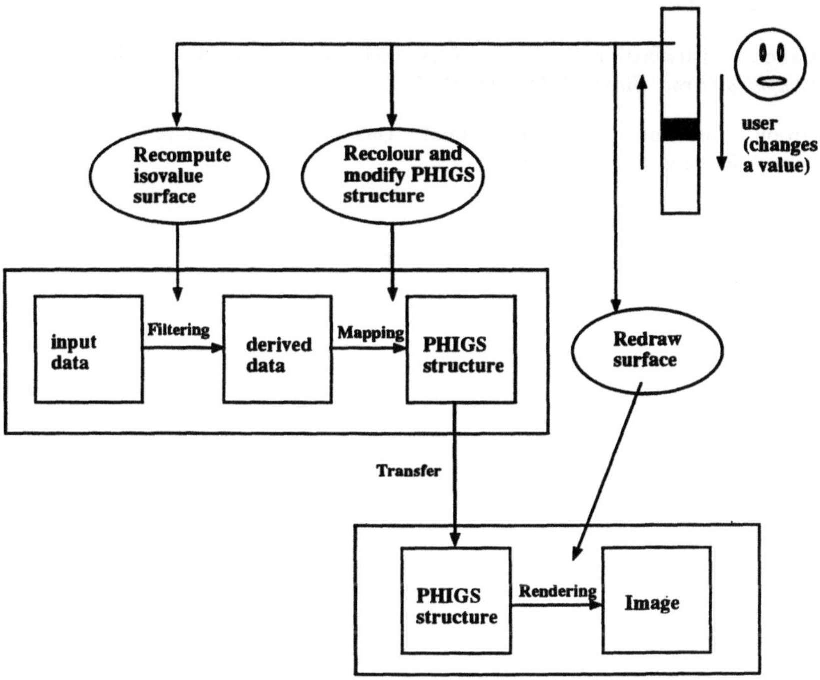

FIGURE 6.4. Visualization of an isovalue surface with a moving value

We can see how the system works (figure 6.4) in the case of architecture 3. We have supposed that the user is interested in isovalue surfaces and wants to see how a such a surface is deformed when its corresponding value changes. The key point is the transfer of data from the Cray computer to the graphics engine of the workstation. Not only is the network a potential bottleneck so is the cpu of the workstation. One must avoid decoding the data arriving from the network to store them in another format at the input of the rendering pipe-line. A DMA access from the Cray to the workstation could provide a good solution. If we want to be able to visualize the ongoing deformation of a surface according to the value entered by the user then the response time of the system must be as short as possible. This justifies the utilization of a supercomputer, a 3D-graphics workstation and a high speed network.

6.5 Conclusion

We have pointed out that scientific visualization requires very powerful systems especially in fluid dynamics where animation is necessary to visualize a fluid flow.correctly. A single workstation can be used but one finds compute intensive functions in the visualisation tools that a workstation cannot process fast enough. One can improve the system by using a supercomputer which will cooperate with the workstation through an adequate network. In the Iridium project, we propose to experiment with different cooperative architectures to implement a post-processor with a high level of performance.

6.6 References

[1] Haber. Visualization in Engineering Techniques, Systems and Issues. *Computer Graphics (Proc. Siggraph 88)*, 22, 1988.

[2] Upson. Two and Three Dimensional Visualization Workshop. *Computer Graphics (Proc. Siggraph 89)*, 23, 1989.

7 Towards a Reference Model for Scientific Visualization Systems

W. Felger, M. Frühauf, M. Göbel, R. Gnatz, G.R. Hofmann

ABSTRACT
Reference models have been developed in various fields of information processing. The aim of such models is to define a unique basis for system development, system usage and for education and training.
One of the first reference models in Computer Graphics was developed by Guedj et al[9]. Today, a (Standard) Computer Graphics Reference Model is under development by ISO[1]. In areas like CAD reference models have already been established on a national basis[4]. The ermerging Imaging standard[2] also defines a reference model for image operations.
In *Scientific Visualization* various system models have been presented in recent years. These models focus on different aspects, such as a model for the visualization process, error accumulation, output pipelines in the visualization process, semantics of interaction in the visualization process, architecture and hierarchy of software modules, computing architectures and load sharing models, data and image interfaces.
These models mostly have been set up by users than by developers of tools. Each of the above models does not reflect the meaning of scientific visualization for it's own but just a certain view. Existing standards, like those known from Computer Graphics (GKS, PHIGS, ...) are not covered by these models.
This paper introduces a reference model for visualization systems and classifies existing models using the criteria of this reference model as a basis for comparison.

7.1 Introduction

Scientific visualization evolved to play a more and more important role in scientific computation. It gives researchers the opportunity to explore their data with the aid of Computer Graphics[12]. Scientific visualization offers methods for seeing the unseen. Symbolic or numerical data is mapped to geometric data with graphical attributes. This data finally is transformed in graphical information with the aim to give insight to the scientist. All these steps should be carried out under interactive control of the user. Thus not only the results of the computations but the simulation, computation and visualization process itself have to be visualized. Since images are produced with the aim to give insight, the semantic of the graphical representation is essential in visualization. Not every mapping of numerical data to geometric primitives may be useful for exploring the data set in the best way. Thus a tool box of transformations from data to images, image to data, data to data and image to image have to be provided by a scientific visualization system.

Reference models as defined in various fields of information processing intend

- to establish a conceptual framework in which definite visualization systems can be classified and compared,
- to describe the major characteristics of visualization systems,
- to define an open visualization system that allows the modification and enhancement of existing functions,

	Image Data	Symbolic Description
Image Data	image processing	computer graphics
Symbolic Description	computer vision	other computational domains

TABLE 7.1. Classification of Computational Domains

- to advance the description, the comprehensiveness and the exchange of the ideas in scientific visualization,
- to install a consistent and overall accepted terminology,
- to identify needs for computer graphics standards, data interchange formats and external interfaces,
- to install a basis for the development of future standards in scientific visualization.

More formally, a reference model is a *set of properties* which may apply to components in visualization systems. Given such a *set of properties* one may check which *properties* are fulfilled by a specific visualization system. This allows the analysis and comparison of visualization systems which differ from applications using other algorithms (so called "check list approach").

Depending on the current interest a user may have, different sets of properties may be taken into consideration. In this way, a particular set defines the *view of interest.* In principle, each discussion, analysis or comparison of visualization systems needs to previously agree on the supposed *view of interest.*

The following discussion on a reference model was heavily influenced by previous work, such as the computer graphics reference model (CGRM)[1], the CAD reference model[4], the imaging standard work item proposal[2], and also current R&D activities at TU-Munich[8] and at FhG-AGD in Darmstadt[6, 5].

7.2 Fundamentals

Scientific visualization comprises techniques from various computational domains such as imaging, computer graphics, computer vision, etc. As visualization also deals with the relationship between image and non-image data, the model in table 7.1 could be considered as a reference. Unfortunately, this diagram does not describe the visualization process in an exhaustive manner and needs further study and clarification.

The fundamental issue in visualization is the *notion of image.* Alternative notions are as follows:

- Painting is a "representation upon a surface of 'visual reality' through colours"[3]. This is derived from human impression, resp. human craftmanship.
- More formally, an image is a function which associates each point of the output area (image plane) with a colour. This mathematical model of the notion of image describes such a function by

$$\mathbf{P} : [R^2 \rightarrow C], \tag{7.1}$$

where R is the set of real numbers and C an appropriate set of colours.

- Another image model for images $\mathbf{P}$ can be seen as

$$\mathbf{P} \subseteq [Z^n \times T \times C^m] \tag{7.2}$$

with

- Z^n, $n = 2$, $n = 3$: spatial discretation (image plane, image space)
- T: time, $T \in Z^+$
- m: number of colour channels (e.g., $m = 1$ for grey levels, $m = 3$ for RGB)
- C^m: colour space, $C \in Z$: discrete amplitudinal value
- $\mathbf{P}$: is a spatial, temporal, amplitduinal and channel discrete structure.

In general we assume only the existence of a *picture domain* P which serves as the *graphical semantical domain* for further formal considerations. The above models (b) and (c) are examples for this picture domain. A particular and detailed specification of the picture domain depends on the *view of interest.* Each data structure m (within a set of data structures) may be associated with (some) graphical semantics by *semantical mapping*; α [10, 14, 7];

$$\alpha : m \rightarrow P \tag{7.3}$$

A set m of data structures associated with a graphical interpretation α (*graphical semantics*) is called a set of *graphical data structures.* Each set m' of data structures that is mapped into m by a function β inherits the graphical interpretation α' in the following way:

$$\alpha' = \alpha \bullet \beta \tag{7.4}$$

$$\begin{array}{ccc} m' & \stackrel{b}{\rightarrow} & m \\ \alpha' & \searrow & \downarrow \\ & & \mathbf{P} \end{array} \tag{7.5}$$

This is an expression for the basic principle of building graphical output pipelines:

$$\begin{array}{l} m_n \stackrel{\beta_n}{\rightarrow} m_{n-1} \stackrel{\beta_{n-1}}{\rightarrow} \ldots m_3 \stackrel{\beta_3}{\rightarrow} m_2 \stackrel{\beta_2}{\rightarrow} m_1 \stackrel{\beta_1}{\rightarrow} \quad m \\ \hfill \downarrow \alpha \\ \hfill \mathbf{P} \end{array} \tag{7.6}$$

Note, that sometimes the set m_i is given as the cartesian product of several sets m_i^j such that:

$$m_i = m_i^1 \quad \times \quad m_i^2 \quad \times \quad \ldots \quad \times \quad m_i^n \tag{7.7}$$

These components m_i^j of m_i are usually known as attributes of graphical data structures, e.g., geometric attributes, colour, texture, etc. In a particular visualization system these attributes may be assigned seperately to each stage of the pipeline! For a given data structure m_i several graphical interpretation functions may coexist! Moreover, applying the same methodology, other types of interpretation functions may be given, such as speech and sound (audio), tactile input, product definition data among other information

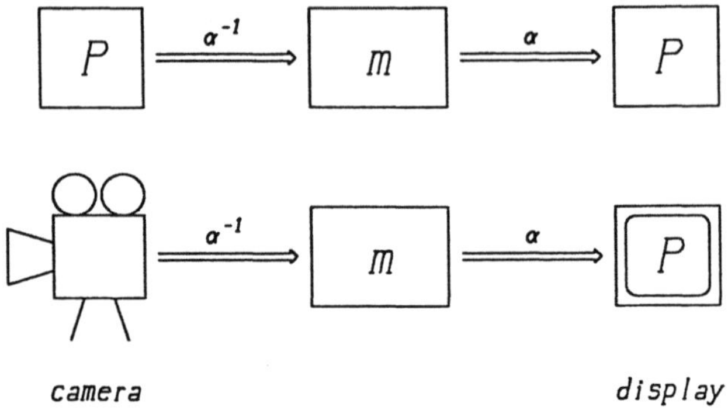

FIGURE 7.1. Semantical mapping α (a: in general, b: specific)

types in the multi-media-context. The graphical interpretation a is usually not a one-to-one-mapping, i.e., there may be different elements in m which produce the same image. Consequently, the inverse

$$\alpha^{-1}(p), p \in P \tag{7.8}$$

is a subset of m — if the inverse exists at all.

Note, however, that the functions β may be transformed according to the composition of components m_i^j (see equation 7.7) such that the different components may be handled seperately in the sense of attribute setting, deffered attribute binding, etc. We assume that within a visualization system the functions $\beta_n \ldots \beta_1$ (see equation 7.6) are implemented by concurrent processes. A general consideration of these aspects leads to a bipartite graph representing the visualization system.

7.3 The basic model

Summarizing the fundamental discussion from above, the *basic model* for visualization systems is set up: figure 7.3 examplifies that data structures of M, with

$$M = \bigcup_{i=0}^{k} m_i \tag{7.9}$$

have a graphical interpretation *alpha* over some graphical semantical domain P. $P*$ is the subjective impression/imagination of P as realized by the "intellectual power" of the user. This means that visualization supposes a correspondence between P and $P*$ illustrated by $P \approx P*$. The arrows illustrate the data flow. Data is reveived from input devices (I), manipulated by processes (pr) and sent to output devices (O).

The basic model represents a projection of the bipartit graph (see figure 7.2). The nodes of type "storage" are mapped onto node M in the basic model. Nodes of type "process" are classified into processes which are *data sources*, *data absorbers*, *data transformers* and other processes. Representatives of data sources are input devices and sensors. Data absorbers are typically output devices like displays and plotters. Data transformers are processes for data sampling, data enrichment and enhancement, data-to-geometry transformations, rendering processes, image-to-image transformations, etc. Other processes comprise configuration tasks, system modification and administration and user interface

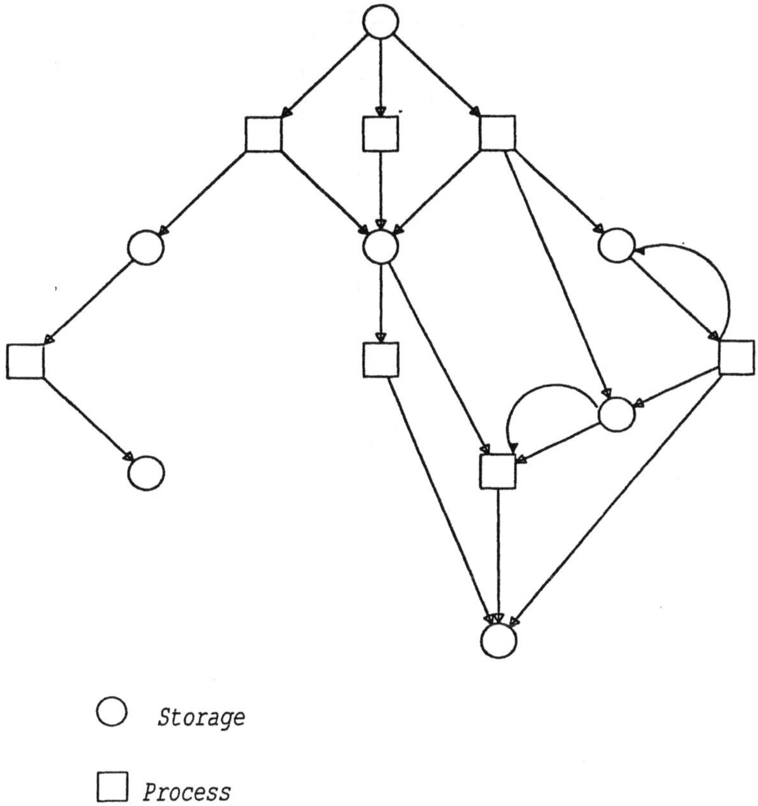

FIGURE 7.2. A bipartit graph representing a visualization system

support including documentation and online help. Now, the basic reference model for visualization systems, may be seen under different *views of interest*, namely

- data flow
- control flow
- graphical semantics
- device abstraction
- system administration and
- user interface.

It is supposed that the basic model is the intersection ("common dominator") of all views of interest.

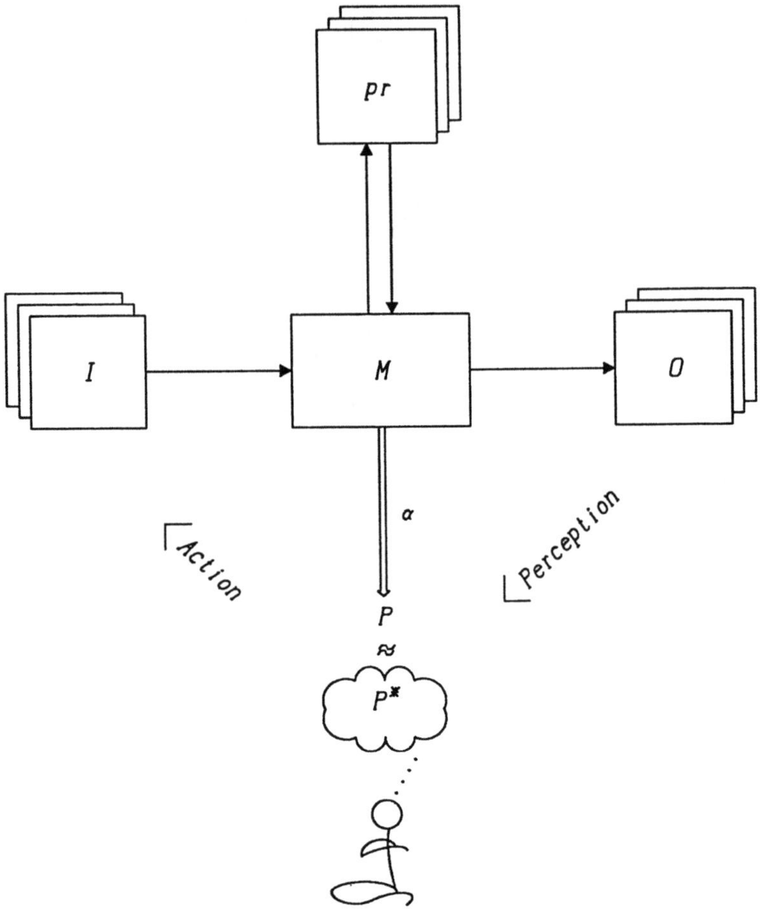

FIGURE 7.3. Basic model for visualization systems

7.4 Derived and detailed models

Given the basic model associated with a specific set B of properties (the basic view B). It is possible to derive from B a new view B' by adding new properties of interest. B' then becomes the *view of interest*. In particular, it becomes obvious that in principle there is a whole lattice of reference models with the basic model as its bottom element.

7.4.1 Modeling in scientific visualization

The emphasis of this model is put on the tranformation of physical reality to a numerical solution of simulations. Different layers in a hierarchy of abstractions from computation are defined. The part of computer graphics in visualization is compressed to the analysis box at the bottom of the hierarchy. The steps below, beginning at 'Mathemaical Formulations of Laws' are the tasks of todays visualization systems. A more abstract modelling can only be performed by the aid of expert systems or artificial intelligence. Nevertheless, the

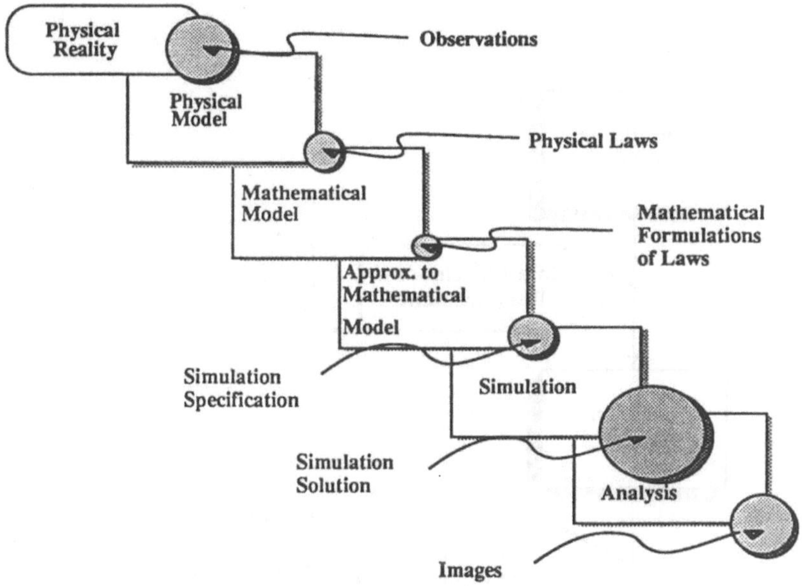

FIGURE 7.4. Modeling in Scientific Visualization[13]

formulation of physical reality by means of programming languages is one of the most challenging tasks in scientific visualization. This task is of course application dependent. The circles shown in figure 7.4 correspond to the nodes of type 'data' in our basic model (figure 7.3). The boxes indicating processes correspond to our nodes of type 'process'.

7.4.2 The visualization process model

This model defines how data is converted from 'row' simulation data to digital images using various functions. The first phase in the pipeline operates on data. Applying interpolations or filters the data is enhanced, enriched or even completed before a mapping from symbolic to geometric data is performed. Different geometric objects may be choosen for visualization. The final phase is the image generating phase. A particular rendering technique may be selected and the result of this process, the digital image, may be varied by specifying rendering parameters. This model shows a pure output pipeline for the visualization process. Visualization is seen only as post-processing of computed data. The user is just passive viewer of the visual results. Various data sets M are shown (simulation data, derived data, abstract visualization objects and the displayable image). The functions β, realized by processes pr are explicitely named, e.g., data enrichment processes or rendering processes. This model classifies the functions β according to their appearance in the output pipeline.

7.4.3 The imaging model

In the ISO/IEC JTC1 SC24 New Work Item Proposal on Imaging[2], there is a diagram for the Overall System Environment, which may be seen as a reference model for the new image processing and interchange standard. The imaging applications will deal will digital

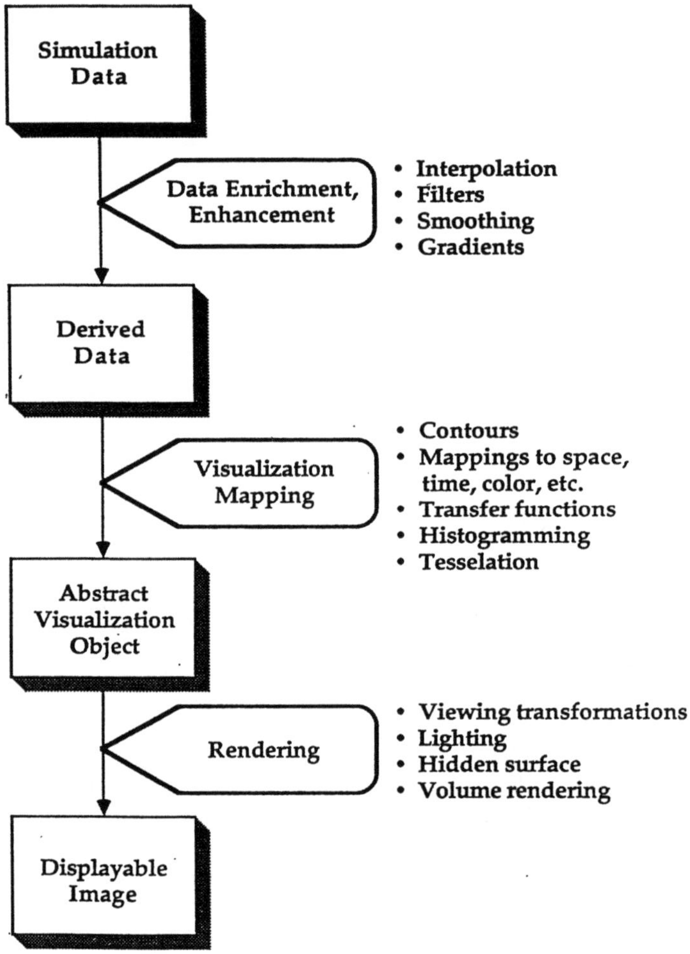

FIGURE 7.5. Visualization Process Model[11]

images. Thus, in the diagram, the digital image (or digital image sequence) is placed in the centre. The three main information or data types are: digital image, image related information (such as numeric measurements, and real world information). As categories of functions, we have sense, display, image-to-image transformation analyze, synthesize, visualize and interpret. The distiction of two functions is obvious by their data inputs and outputs. Three loops may be identified in the diagram: the image-to-image loop, the sense and display loop, and the analysis and synthesis loop. User interaction may be required in each of them.

1. The image-to-image loop is for the direct manipulation and enhancement of the digital images and transforms them in a useful manner.

2. The sense and display loop consists of sense and display for transacting real world information into digital images, and vice-versa.

3. The analysis and synthesis loop show the relationship between computer graphics and image evaluation.

An important work item for imaging involves the exchange of digital images between applications. Image interchnage is not a transducing function, but shares the same underlying data model, as used for other functions in the imaging application programmers interface.
With respect to the basic model for visualization systems (as shown in figure 7.3), we have all data types in the data type pool M, as we have all functions (either transducing or interchange functions) in the process pool pr. The sense and display loop of the imaging diagram correspond to the I and O items of the Basic Model. The user, however, is dealing with the real world information of the diagram, the user is not explicitely mentioned and drawn in the imaging diagram.

7.4.4 The visualization taxonomy

This model focusses on the loops in visualization systems. The architecture of visualization systems can not be simplified to a pipeline structure, since user interaction demands loops in certain steps of the visualization process. This model distinguishes between two types of data. These types are image data and non-image data. Data is transformed by algorithms of image from image processing, computer vision, computer graphics and other computational areas (as shown in figure 7.1). Processes from our basis model correspond to the arrows in this model. Data is separated in two types and examples for input and output devices are explicitely introduced.

7.5 Conclusion

We have set up a basic model for comparison and analysis of visualization systems in scientific computing. Existing models in visualization have been discussed using this basic model. Moreover, this basis model is valid in other domains, like imaging and CAD. Details built into the basic model lead to models for specific visualization systems, such as presented in section 7.4.

Acknowledgements:

This work was granted by the German Ministry for Research and Technology (Bundesministerium für Forschung und Technologie, BMFT). The work was coordinated by the German Computer Science Society (GI).

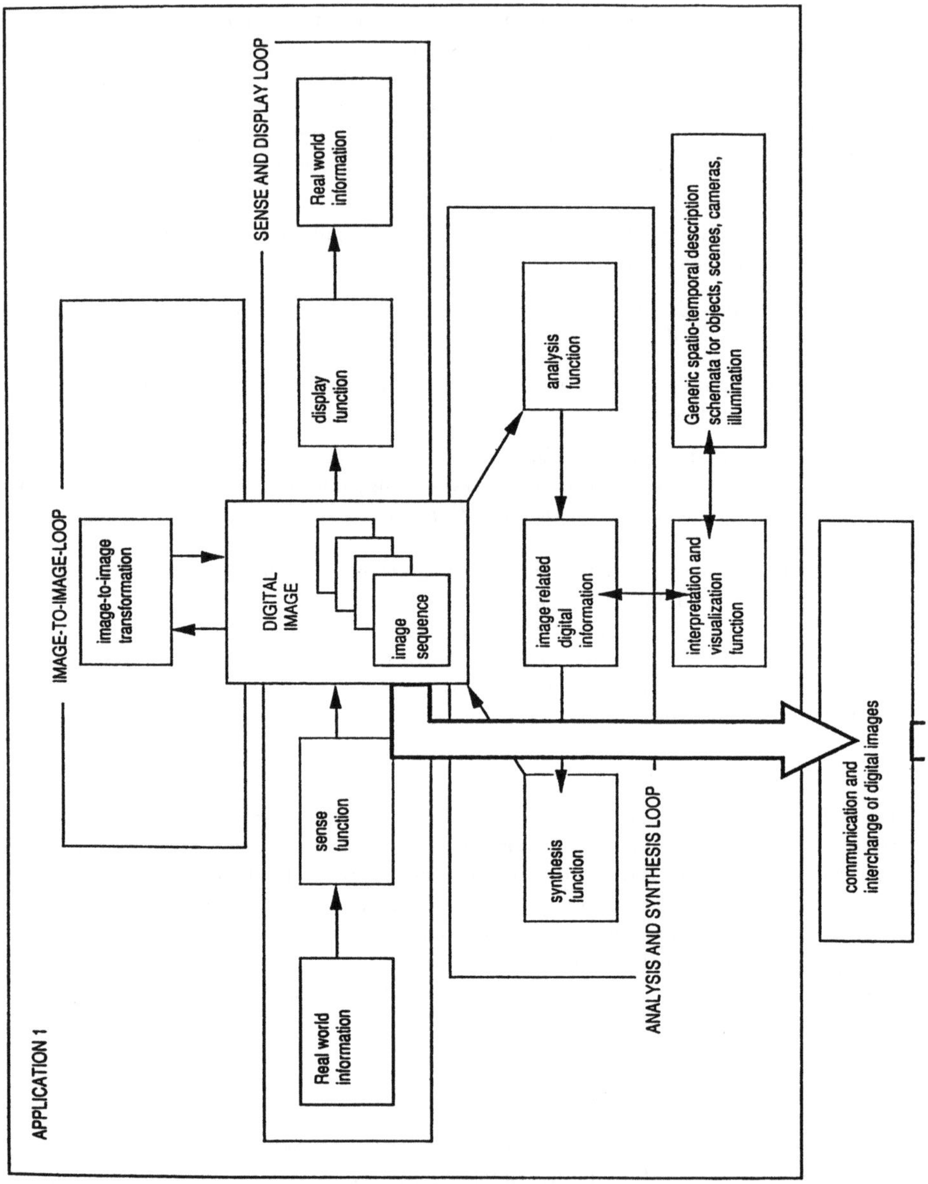

FIGURE 7.6. The Imaging Model[2]

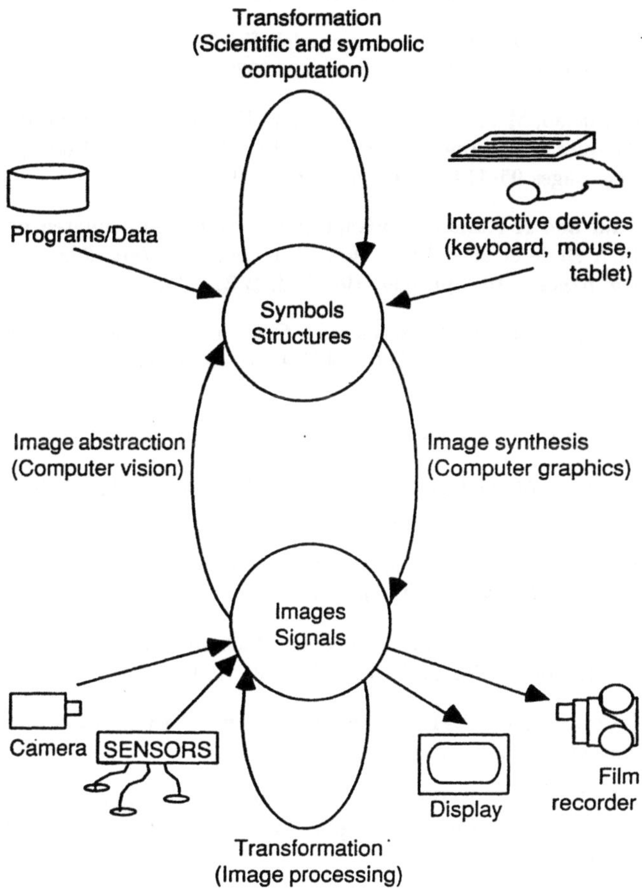

FIGURE 7.7. The Visualization Taxonomy[12]

7.6 References

[1] *ISO/IEC JTC1/SC24/WG1: Computer Graphics Reference Model*, 1990.

[2] *ISO/IEC JTC1/SC24/WG1: New Work Item Proposal - IMAGING: Image Processing and Interchange Standard*, 1990.

[3] S Dali. *Dali. 50 secrets magiques.* Lausanne, Paris, 1985.

[4] DIN. GI-FG 4.2.1, AK1: Referenzmodell für CAD-Systeme, Gesellschaft für Informatik. Technical Report, DIN, February 1989.

[5] J L Encarnacao, M Frühauf, M Göbel, and K Karlsson. Advanced Computer Graphics Techniques for Volume Visualization. In H Hagen and D Roller, editors, *Geometric Modelling*, pages 95–114. Springer, Berlin, 1991.

[6] M Frühauf and K Karlsson. Visualisierung von Volumendaten in Verteilten Systemen. In A Bode et al., editors, *Visualisierung von Umweltdaten in Supercomputersystemen*, pages 1–10. Springer, Berlin, IFB 230, 1990.

[7] R Gnatz. Specification of Interfaces: A Case Study of Data Exchange Languages. In *Product Data Interfaces in CAD-CAM Applications*, pages 192–207. Springer, Berlin, 1986.

[8] R Gnatz. LRZ - Anwender Workshop. Technical Report, TU Munich, April 1989.

[9] R Guedj et al. *Methodology in Computer Graphics, Seillac I.* North-Holland, Amsterdam, The Netherlands, 1976.

[10] J Guttag and J J Horning. Formal Specification as Design Tool. Technical Report CSL-80-1, XEROX-PARC, January 1980.

[11] R B Haber. Visualization in Engineering Mechanics: Techniques, Systems and Issues. Technical Report, ACM SIGGRAPH, 1988. ACM Siggraph'88, Course Notes 19.

[12] B H McCormick, T A DeFanti, and M D Brown. Visualization in Scientific Computing. *Computer Graphics*, 21(6), November 1987.

[13] C Upson, D Kerlick, R Weinberg, and R Wolff. Two and Three Dimensional Visualization Workshop. Technical Report, ACM SIGGRAPH, 1989. ACM Siggraph'89, Course Notes 13.

[14] M Wirsing et al. On Hierarchies of Abstract Data Types. *Acta Informatica*, 20:1–33, 1983.

8 Interactive Scientific Visualisation: A Position Paper

R.J. Hubbold

ABSTRACT
This paper summarises the author's views on current developments in interactive scientific visualisation. It is based on a talk presented at the Eurographics '89 conference, held in Hamburg in September 1989. The paper takes issue with the direction of some current work and identifies areas where new ideas are needed. It has three main sections: data presentation methods, current visualisation system architectures, and a new approach based on parallel processing.

8.1 Introduction

The upsurge of interest in visualisation was given a major impetus by a report prepared for the National Science Foundation in the USA (the ViSC report) [16]. The main thrust of this was to examine how the USA could remain competitive in this area, and therefore what research should be funded by the government. A major problem identified was how researchers could assimilate the truly vast amounts of data being poured out by supercomputers — what the report termed "firehoses of data". The report recommended a specific approach to the structuring of systems for scientific visualisation, largely determined by the view that numerical computing would be performed by supercomputers, which by their nature are very expensive and therefore centralised and shared. Viewing of results and interaction would be done locally, using *visualisation workstations.* This arrangement demands ultra-high speed networks, and the funding of these was one of the report's recommendations.

This separation of graphics and interaction from application computations is a familiar theme in computer graphics. For example, it underpins the design of graphics standards such as GKS [9] and PHIGS [10]. In this paper, it is argued that this approach creates *inflexible* systems which are not appropriate for the purposes of scientific visualisation.

As a starting point it is useful to define the term *visualisation.* The Oxford English Dictionary gives:

Visualize: *to form a mental vision, image, picture of. To construct a visual image in the mind.*

Visualization: *the action, fact or power of visualizing; a picture formed by visualizing.*

and Chambers Dictionary has:

Visualisation: *to make visible, externalise to the eye: to call up a clear visual image of.*

whilst Roget's Thesaurus, under *visualize*, gives:

See/know: *behold, use one's eyes, see true, keep in perspective, perceive, discern, distinguish, make out, pick out, recognize, ken, take in, see at a glance, discover ...*

Imagine: *fancy, dream, excogitate, think of, think up, dream up, make up, devise, invent, originate, create, have an inspiration ...*

These definitions convey a clear meaning: that visualisation is concerned with the formation of *mental images*, or *models* — the notion of "seeing something in the mind's eye". The NSF report correctly identified this aspect of visualisation, and referred to the key goal of providing *insight* into diverse problems. However, it cast the net much wider than this and defined visualisation as the integration of computer graphics, image processing and vision, computer-aided design, signal processing, and user interface studies. It also identified specific applications which might benefit, for example, simulations of fluid flow, and studies of the environment. Unfortunately, many people have chosen a much narrower interpretation, and the term *visualisation* is now frequently abused. Too often, it is used to refer *only* to the synthesis of images, and especially to attempts to generate complex three-dimensional images in near real-time.

In this paper, the term *interactive scientific visualisation* is employed to emphasise the long-term goal of enabling interaction with large-scale numerical simulations — so-called *user-steered calculations*. The remainder of the paper addresses three areas:

1. Techniques for displaying data.
2. Current graphics system architectures and their use for visualisation.
3. Parallel processing and interactive visualisation.

It is suggested that, in the next five to ten years, developments in parallel processing will begin to mature, to the extent that new approaches will not only be possible, but essential if the goal of gaining insight into the behaviour of complex systems via user-steered calculations is to be achieved.

8.2 Display techniques

Many recent developments in display techniques have been driven by the quest for visual realism. Clearly, for some purposes this is a worthy goal — see, for example, recent results from radiosity algorithms and their potential use in architecture [6]. Large numbers of problems, particularly simulations of physical phenomena, deal with situations which evolve over time, so the use of animation techniques seems a logical choice. It has been assumed without much scrutiny that if graphics systems are developed to the point where realistic images can be generated in near real-time then many of the problems of scientific visualisation can be overcome. This overly simplistic view has obscured the need to examine and develop alternative, and sometimes cheaper, methods for data display.

8.2.1 Animation

The use of animation for scientific analysis of complex results is an area fraught with difficulty. Animated sequences can be a wonderful way to convey an impression of behaviour, but are not so valuable for quantitative comparisons.

- Three-dimensional pictures displayed on a flat screen rely on a range of cues to assist the viewer. Amongst these, relative motion of objects at different depths is very popular — hence, in part, the desire to rotate 3D scenes in real-time. Often, when the rotations cease then depth ambiguities appear. Unfortunately, when objects are moving the human visual system cannot track fine detail in the picture. Thus, systems which generate very realistic scenes may be doing an unnecessary amount of computation. Conversely, if a coarse picture is employed which contains certain kinds of artifacts, such as aliasing effects, then, perversely, the viewer's attention seems drawn to the defects which may be exagerated by animation.

- A particular problem with animation is that it does not permit easy comparison of different frames. Techniques are needed which facilitate the display of different time steps, either side by side or superimposed, with transparency techniques and colouring schemes employed to highlight differences.

- One way to display motion in a continuum is to use particle clouds. Upson et al [19] report that motions of individual points can be tracked if the number of particles is small, but that as the point density is increased then ambiguities occur. This is a form of *temporal aliasing*, in which different points become confused between frames, so that points may even appear to move in the wrong direction — the waggon wheel effect familiar in old films. As the number of particles increases still further the authors report that cloud-like motions can be observed.

8.2.2 Visual interpretation

- So much effort has gone into simulating realistic lighting that some more fundamental aspects of deducing shape from shading may well have been overlooked. In a particularly interesting paper [17], Ramachandran argues that humans are conditioned to seeing objects illuminated from above. He shows convincingly that the human visual and cognitive system is able to discern patterns in test pictures where objects are assumed to be lit from above, which are simply not visible if the objects are lit from the side.

- It is not obvious how to display multi-dimensional data. Experiments on this are progressing at a number of places. The NCSA in Illinois has produced a stunning example of three-dimensional animation showing the development of a major storm system [5]. Their display uses a variety of techniques, such as transparent surfaces, symbols, ribbons, arrows and colours to show a total of nine different dimensions in the model. Significant effort has been devoted to finding new ways to show all of these factors, including the involvement of artists. The results have been produced using a fully equipped video recording studio; it is not possible to interact with the model directly. Notwithstanding such efforts, the use of three-dimensional techniques to display such results is still in its infancy, and users require training before they can interpret the results.

- Colour has no intuitively obvious interpretation, except that blue is usually regarded as cold (low) whilst red is hot (high). Some experts [14] advocate using the spectral order:

  (low) V I B G Y O R (high)

  to show a range of values. But, if we heat a metal bar we know that the colour changes from red, to orange, to yellow, to white as temperature increases — that is, in the reverse of the spectral order!

8.2.3 The display of quantitative information

Traditionally, many scientists make extensive use of graphs and charts. Several companies which market software for displaying information in this form are capitalising on interest in visualisation; see for example [11]. This type of presentation requires careful thought.

In his excellent book Tufte [18] gives numerous examples of good and bad practice for the display of quantitative information. Computer-generated figures come in for some justified criticism. Indeed, it is difficult to see how some of the graphics in his book could be produced by a program without considerable difficulty. An example is the well-known map by Minard, depicting the advance of Napoleon's army on Moscow in 1812, which not only conveys the geography of the situation but contains a wealth of statistical information.

The book contains a panoply of methods for data display which are potentially useful for visualisation of quantitative data, and especially for time-dependent phenomena. Examples include mixed charts and graphs, the use of tables, and rugplots. A common timebase for a set of graphs may well reveal dependencies between parameters in a model which are not evident in animation sequences. This has the distinct advantage that results may be studied carefully and in detail, since they do not change before the viewer's eyes!

These simple methods do not generate the excitement of three-dimensional animation, but they are nonetheless properly a part of scientific visualisation. More work on such techniques would certainly be warranted, especially on methods for showing relationships and dependencies between parameters.

8.3 Current visualisation system architectures

The NSF ViSC report outlined an approach to visualisation which assumed that large-scale computations would be carried out on supercomputers and display and interaction would use visualisation workstations, with the two connected by very high-speed networks. This is very much the strategy currently being pursued by researchers and by some hardware vendors. There is major interest in network computing which aims to put the display and interaction on the user's desktop, and developments such as the X Window System [12] and PEX [3] reinforce this direction.

One clear benefit of the separation of the graphics from other parts of the application is that specialised hardware can be designed to support the computationally-intensive parts of the graphics pipeline, especially transformations, clipping and rendering. It also permits, to some degree, the definition of device-independent graphics systems, such as PHIGS PLUS. However, there are problems with this policy which have been recognised for many years:

- It becomes necessary to have two representations for the problem data, one graphical and one application-oriented. The specialised data structures of systems such as PHIGS (and PEX, by implication) are useless for many application tasks. Application programs are thus forced to duplicate information and to provide algorithms for mapping each representation on to the other and for updating both in a synchronised manner.

- In a network environment a decision must be made about what tasks to perform locally and what to do remotely. This thorny issue, identified as long ago as 1968 by Myer and Sutherland [15], has plagued system implementors for years. Any solution tends to be dominated by current hardware. As technology changes, the balance of processing for the ideal solution keeps migrating back and forth between the remote and local processors.

- Specialised graphics workstations tend to achieve high performance by casting very specific algorithms and data structures into hardware or microcode. These are virtually "black boxes" to the application programmer and are inflexible in the sense that the end user cannot re-program the device to do anything differently. For example, PHIGS uses a particular data structure which precludes the use of multiple inheritance. An important aspect of scientific visualisation is the need to explore new methods of presenting data, which requires flexible programmable systems.

- Many current systems are heavily dependent on using polygons for graphical modelling. It is far from clear that polygons are an appropriate way to model certain problems, especially those where the model may change significantly between frames. Curved surfaces can require huge numbers of polygons for a reasonable approximation. To take a simple example, a decent rendering of a smallish sphere requires anything up to 1000 triangles. At this rate, systems which can render tens, or even hundreds, of thousands of triangles per second very soon begin to struggle when asked to display a large number of spheres. Fortunately, some displays are able to scan-convert spheres and other quadric primitives directly, but other representation techniques are badly needed.
- Related to this issue is that of creating the polygon-based data. Most vendors quote rendering times for their hardware which assume that this representation already exists. The time to generate a data format acceptable to the display system is frequently one to two orders of magnitude slower than the raw rendering speed. This becomes a severe problem if the model can change significantly between frames. Anyone who requires convincing of this should compare structure generation times required by PHIGS implementations with the corresponding traversal and rendering times. (PHIGS data structures are complicated! [8])
- In future, user-steered calculations will require improved user interfaces which permit the operator to interact more closely with application models. The aim should be to achieve near real-time *semantic* feedback, rather than simple, local input device echoing (lexical feedback). Semantic feedback is generated as a result of application computations — new results and constraint checks — whereas lexical echoing takes the form of highlighting menu choices, cursor tracking and other similar, low-level techniques. In a distributed environment, semantic feedback requires round-trips between the user's workstation and the computation server. In the author's view, a tighter coupling between the application processing and user interface compnents of a system will be necessary than is common today, and this is not merely a question of providing higher-speed networks. (As an aside: the X Window System requires round trips even for simple lexical feedback. The problems of round-trip delays are likely to become evident as X11 is more widely used.)

8.3.1 The graphics pipeline

A useful way to characterise the graphics pipeline, proposed by Akeley [1], is to divide it into five major sections: G-T-X-S-D:

G is for Generation. It is concerned with the definition of graphic primitives, attributes and their structure in a format acceptable to the remainder of the pipeline. In a typical system, such as PHIGS, this would comprise the generation of the structures, primitives, attributes and transformations.

T is for Traversal, and deals with the flattening of any graphics model structure into a sequence of primitives prior to individual transformation and display. It caters, for example, for instancing mechanisms.

X is for transformation. It is usual to employ homogeneous coordinates in which transformations can be represented by square matrices. It then becomes possible to design an implementation in such a way that all transformations can be concatenated to yield a single composite matrix. This applies even to systems such as PHIGS, which has a large number of diferent transformation and clipping stages. [7]. However, care is needed to make sure that primitives are perspective invariant. This is true for lines, polygons and NURBS (hence the interest in the last of these).

S is for Scan-conversion. This is the stage which is concerned with mapping a geometric definition into an image. It includes the conversion of geometric data into pixels and the application of lighting, shading, and texture mapping algorithms to produce shaded displays.

D is for Display. It is concerned with image storage and with compositing techniques such as α-blending and z-buffer hidden surface removal. The boundary between the **S** and **D** stages is sometimes indistinct.

Hitherto, many workstations have implemented all stages of this process, except generation, in special hardware. They can be characterised as **G-TXSD**[1] systems. There are already indications that display manufacturers have begun to recognise the dangers of this "black-box" approach. Machines like the Ardent Titan [2] and Apollo DN10000 [20] carry out much of the graphics pipeline processing in their general purpose processors, rather than in special hardware, yielding greater flexibility. The Titan can be characterised as a **GTX-SD** machine, in which only the scan-conversion (of triangles) and display are handled by special hardware. Arguments against the **G-TXSD** configuration are presented very cogently by Akeley [1].

8.4 Parallel processing and interactive visualisation

Within five to ten years many large-scale simulations will be carried out on parallel computer systems. Parallel architectures will offer an opportunity to develop radically new solutions to problems, and very different visualisation methods and system architectures. A key component of future interactive visualisation systems will be the integration of visualisation and application computations.

Thus far, attempts to use parallel processing for display have centred on mapping the graphics pipeline on to several processors, using multiple instruction, multiple data stream (MIMD) architectures, or on using single instruction, multiple data stream (SIMD) systems to perform scan-conversion. There have also been several implementations of ray tracing, which integrates the different stages of the pipeline into a single algorithm. A good survey of alternative approaches can be found in [4] and [13].

Significantly, relatively little work seems to be in progress on examining new graphics architectures, in which application computations and visualisation algorithms are closely coupled and distributed over large numbers of processors.

Deriving new solutions to visualisation tasks for parallel machines is a challenging problem. Current parallel architectures tend to be good at parts of the process, but not all of it, because image synthesis involves large amounts of computation *and* large amounts of data. MIMD systems are often good at computation but suffer data bandwidth problems (e.g., transputers), whilst SIMD systems are good at certain image-level operations but poor at

[1] Stages to the left of the hyphen are performed by general purpose, user-programmable CPUs, and those to its right are embedded in specialised graphics hardware.

higher-level tasks. Sometimes they too suffer data bandwidth problems (e.g., broadcasting data on a Connection Machine), or processor utilisation problems when performing image-space calculations on small objects. Shared memory (shared bus) architectures tend to be limited by the number of processors which can be configured (typically eight). These problems have led some researchers to propose that the eventual solution for graphics may be a hybrid MIMD/SIMD system. (In a small-scale way, some commercial systems already have this feature, employing a SIMD "footprint" engine for polygon filling and shading.)

8.4.1 The PARAGRAPH project

At the University of Manchester we have begun to research a new visualisation system, currently called PARAGRAPH, in which as much as possible of the graphics computation will be performed using large-scale but general purpose parallel architectures. This approach is currently unfashionable because it is generally assumed that specialised graphics processors will always provide a more cost-effective method to perform image generation than will general purpose machines.

It is too early to give a very detailed description, but the following points give some idea of our approach.

- At the heart of our thinking is the design of a new three-dimensional imaging model. A major criticism of current graphics systems is the one-way nature of the graphics pipeline. The end user interacts with his model through the medium of the picture. As more advanced interfaces develop (such as virtual worlds) some method of relating image manipulations back to the application model will be needed. Segment, structure naming and picking schemes used by systems such as GKS and PHIGS are very crude and inadequate. We intend to use information stored at the image level to "reach back" into the application model.

- We are investigating ways in which the image-level data can be used to improve both the image synthesis computations and application computations. Current graphics systems waste huge amounts of processing power performing redundant calculations. A simple example is the non-optimised use of a brute force z-buffer hidden surface algorithm. We are looking at whether it is possible to use refinement techniques in which a quick-pass algorithm can generate image data which is applied subsequently for high-quality image generation. In the longer term we hope that this kind of refinement can be propagated back into the application calculations, so that detailed simulation calculations are only applied to areas of the model which the user is currently exploring. This is a hybrid divide and conquer strategy, applied in 3D image space *and* in object space.

- The 3D imaging model will support a variety of compositing techniques. There are two key aims behind this. First, it will be possible to merge three-dimensional images generated using a variety of different techniques, including CSG operators, and α-blending. We are concentrating particularly on developing a model which permits volume rendering to be integrated with more traditional surface rendering methods. Second, we expect component parts of the image to be computed in parallel and then merged. The use of feedback from the image level will be employed to support lazy evaluation, so that expensive rendering is only performed for parts of the image which are actually visible.

- In conjuction with scientists from a number of application areas we expect to investigate different methods for generating pictorial representations of data. In particular, we will look at alternatives to polygon modelling, although polygon-based models will be supported because they are approprate for some kinds of problem (e.g. finite elements).

- Initially, we expect the application models to be divided in object space and distributed over multiple processors. We intend to examine alternative strategies for distributing the **GTXS** parts of the pipeline, and especially to make the generation and traversal as flexible as possible. We do not expect to be able to implement the whole system without some hardware support, but an aim of the project is to consider carefully just what form this should take. One possibility is to have a **GTXS-MD** system, where **M** stands for Merge, and is closely related to the 3D imaging model.

- As a parallel activity we are examining how to build user interface management tools for parallel systems. In designs such as PHIGS the processing of input is largely divorced from the output pipeline. We intend to link our input tools closely to the imaging model in order to provide support for 3D input, as well as picking and window management.

- Currently, work on development of the imaging model is progressing on an Ardent Titan.

Acknowledgements:

The author is grateful to colleagues in Manchester for helpful discussion, particularly Alex Butler who is currently working on our imaging model.

8.5 References

[1] K. Akeley. The Silicon Graphics 4D/240GTX Superworkstation. *IEEE Computer Graphics and Applications*, July 1989.

[2] B. Borden. Graphics Processing on a Graphics Supercomputer. *IEEE Computer Graphics and Applications*, July 1989.

[3] W.H. Clifford, J.I. McConnell, and J.S. Saltz. The Development of PEX, a 3D graphics extension to X11. In D.A. Duce and P. Jancene, editors, *Proceedings Eurographics '88*. North-Holland, 1988.

[4] P.M. Dew, R.A. Earnshaw, and T.R. Heywood. *Parallel Processing for Computer Vision and Display*. Addison-Wesley, 1989.

[5] R. Wilhelmson et al. Study of a Numerically Modelled Severe Storm. NCSA Video, University of Illinois at Urbana-Champaign, 1990.

[6] D.P. Greenberg. Advances in Global Illumination Algorithms (Invited Lecture). In W. Hansmann, F.R.A. Hopgood, and W.Strasser, editors, *Proceedings Eurographics '89*. North-Holland, 1989.

[7] I. Herman and J. Reviczky. A means to improve the GKS-3D/PHIGS viewing pipeline implementation. In G. Maréchal, editor, *Proceedings Eurographics '87*. North-Holland, 1987.

[8] T. L. J. Howard. A Shareable Centralised Database for KRT^3—a hierarchical graphics system based on PHIGS. In G. Maréchal, editor, *Proceedings Eurographics '87*. North-Holland, 1987.

[9] International Standards Organisation (ISO). *ISO–7942 Information Processing Systems — Computer Graphics — Graphical Kernel System (GKS) Functional Description*, 1985.

[10] International Standards Organisation (ISO). *ISO 9592 Information Processing Systems — Computer Graphics — Programmer's Hierarchical Interactive Graphics System (PHIGS)*, 1989.

[11] M. Jern. Visualization of Scientific Data. In W. Purgathofer and J. Schönhut, editors, *Advances in Computer Graphics V*, EurographicSeminars. Springer-Verlag, 1989.

[12] Oliver Jones. *Introduction to the X Window System*. Prentice-Hall, 1989.

[13] A.A.M. Kuijk and W. Strasser, editors. *Advances in Computer Graphics Hardware II*. EurographicSeminars. Springer-Verlag, 1988.

[14] G.M. Murch. Human Factors of Color Displays. In F.R.A. Hopgood, R.J. Hubbold, and D.A. Duce, editors, *Advances in Computer Graphics II*, EurographicSeminars. Springer-Verlag, 1986.

[15] T.H. Myer and I.E. Sutherland. On the design of display processors. *Comm. ACM*, 11, 1968.

[16] NSF. Visualization in Scientific Computing. *ACM Computer Graphics (Special Issue)*, 21, 1987.

[17] V.S. Ramachandran. Perceiving Shape from Shading. *Scientific American*, August 1988.

[18] Edward R. Tufte. *The Visual Dispay of Quantitative Information.* Graphics Press, Box 430, Cheshire, CT 06410, 1983.

[19] C. Upson, T. Faulhaber Jr., D. Kamins, D. Laidlaw, D. Schlegel, J. Vroom, R. Gurwitz, and A. van Dam. The Application Visualization System: a Computational Environment for Scientific Visualization. *IEEE Computer Graphics and Applications*, July 1989.

[20] D. Voorhies. Reduced-Complexity Graphics. *IEEE Computer Graphics and Applications*, July 1989.

Part III

Applications

9 HIGHEND — A Visualisation System for 3D Data with Special Support for Postprocessing of Fluid Dynamics Data

Hans-Georg Pagendarm

9.1 Introduction

Large and expensive supercomputers are producing an enormous amount of data at significant costs. In order to use these facilities efficiently dedicated peripheral hardware and software is necessary. Computer graphics help the researcher to prepare data for the supercomputer and to process the data produced by numerical solvers.

Data visualization has become a very important topic for many researchers. But it is observed that a large effort is spent for work under this topic by people needing visualization for their main research work. As many visualization techniques are of general use, the idea of a multiple purpose visualization software has come up at many places. Generalized visualization tools also become very desirable, because the major part of visualization techniques is useful independent of a certain application. Nevertheless there is always a part remaining, where destinct knowledge of the appli- cation is necessary.

Two years ago there was no visualization software available on the market, which fitted the needs of aerodynamicist processing their large 3D data. In order to fill this gap the Institute for Theoretical Fluid Dynamics of the DLR in Gottingen decided to design a complex software system for visualization of 3D data sets. As the visualization process itself was recognized to be independant of the aerodynamic problem, a modular concept was chosen. The system consists of a number of tools, some of which deal with special aerodynamics data processing, others perform visualization of graphical objects only. Thus the application dependent parts of the system are well separated from the visualizing modules. A comfortable user interface was build up using a window system and window toolkit. The common user interfaces also integrates all the modules into one single system with unified look and feel. The user is also kept free from keeping track of the organisation of his data by implementing a common data management and data access strategy into all the modules. The system allows a highly interactive style of working, featuring interactive 3D rotation and manipulation of display layouts.

To summarize these properties the system was named **H**ighend **I**nteractive **G**raphics using **H**ierarchical **E**xperimental or **N**umerical **D**ata (HIGHEND). The system supports:

- data structuring and management
- processing of 3D and 2D structured data
- processing of multi-domain 3D block structured data
- processing of 3D unstructured surface data
- calculation of aerodynamic quantities
- 3D interactive rotation with mouse input
- re-usable layout definitions
- colour-coded scalar quantities
- grids and shaded and illuminated surfaces

- display of vector quantities using arrows
- iso-lines
- positioning of streamlines or trajectories by graphical input
- clculation and display of streamlines or trajectories
- conversion of data formats
- combination of graphical objects
- interactive manipulation of all processes and graphical parameters
- open interfaces

While still being extended the system is already used for postprocessing in various aerodynamic research projects and is now suggested to become a standard visualization tool of the fluid mechanics division of the DLR. Therefore it will be ported to various hardware platforms trying to keep open access to special high performance capabilities of this hardware. The present version in use is running on the family of Sun workstations. It is growing rapidly. At present about 3 Mbytes of source code have been written.

It is necessary that the system runs on low priced workstation during the design phase. Nevertheless it is expected that more high-speed workstations will be installed. The system has to run on those as well. Interactive parts of the system must, however, give a still reasonable performance on slow desk top machines.

9.2 Internal design of HIGHEND

When designing graphic systems, the software has to be well suited to the hardware. Modern workstations still need their software designed specially for them in order to get maximum graphic performance. But other criteria play an important role as well. Network throughput, disk space, memory size, size of the datasets to be analyzed, typical operations to be performed on that data, all this strongly influences the performance and therefore should be taken into account, when thinking about graphics.

When fluid dynamic problems have been computed on a supercomputer or when large sets of experimental data have been created in a windtunnel, these data mostly will be processed to obtain a set of graphics to show the significant results. In order to achieve this, a massive reduction of data has to take place. Most of the data will not be published or kept in the archives. Usually only very little meaningful data survives the reduction and analysis phase.

There is no straight forward way to reduce data. Data reduction is extremely problem dependent. This is the reason why many researchers tend to have their own data reduction and graphic tools. In order to offer one single data reduction or graphic system to a variety of research problems, this system has to be very flexible. Such flexibility can be achieved by allowing interactive influence on the process of data reduction and graphic display.

9.2.1 External factors which influence software design

Many factors may and essentially should influence software design. This may be hardware features as well as other environmental factors like the network or the operating system of the computer, where a graphic system should be implemented. Some of these factors will be discussed in more detail to give an insight into the reasons, why a system is designed in a certain way, why a certain system may have disadvantages in one environment while being well suited to a different site.

Very often graphic postprocessing is not done on the usual mainframe computers but rather on special hardware, so called graphic workstations. These workstations sometimes have a dedicated graphic processor to speed up the display. Some of these even perform transformations in 3D as well as other high level graphic functions.

Disk access time is a significant limiting factor as datasets in fluid dynamics tend to be rather large. Often workstations are equiped with inexpensive but slow harddisks. Mainframes may do better there. This advantage can be useful only if the workstation is connected to the mainframe through a very high speed network and if the mainframe is not too busy. Mostly it will be best to store data in the workstation itself. Too small size of main memory may lead to swapping of code and data and thereby put additional load on disks or network.

Some of the limiting factors may be minimized by analyzing the data flow in your site and configuring mainframes, fileservers and graphic workstations in the proper way. Some may be overcome by using certain features of the operating system or other software which is installed on the computer. Using special hardware or software features often means less portability.

Last but not least, it seems to be essential to consider how much effort can be put into maintenance of the software or extension of the software. If the software is going to be maintained by a special staff, it is no problem to use very fancy "hacks" to gain performance. On the other hand, if the software is likely to be spread among researcher, who extend it themselves, it may likely be a problem to build a very complex software.

Influence of data size

In general, graphic workstations will be installed in a network. The network will supply access to a number cruncher for the calculation of flows. The data will then be brought to further processing to the dedicated graphic machines. Naturally the performance of the whole system will be influenced by all components. It is important to know how large the data flow from the number cruncher to the workstations is, how much time the network takes to do this transfer. Depending on the size of the data sets, the performance of the mainframe and the network throughput, different setups for the system hard- and software may be optimal.

Illustrated in figure 9.1 is a typical operations performed on fluid mechanics data. Fluid mechanic data often comes as blocks. Such a block is considered to be an array of floating point numbers ordered along three indices. A number of these blocks aligned form one data set. The blocks may represent quantities like x,y and z coordinates or velocity components or other quantities like pressure, density, Machnumber, vorticity and many more. Usually there is no need to deal with all these quantities at the same time. So the large data set, which is convenient for data transfer and archive purposes may be divided into single blocks, each of which contains only one quantity or one component of a vector or one component of a coordinate but for all possible indices inside the domain. These blocks will be referred to as 3D-matrices from now on. Dividing data in such a way has the advantage, that one has to deal only with that part of the data, which is necessary for the graphics. All quantities can be dealt with, in the same way. For most graphic purposes these 3D-matrices are not the minimum amount of data. Very often only slices of data are needed. A slice is considered to be a subset of the data for which one of the three indices of the 3D-matrix is constant. There may be slices for constant i,j or k index. Obviously

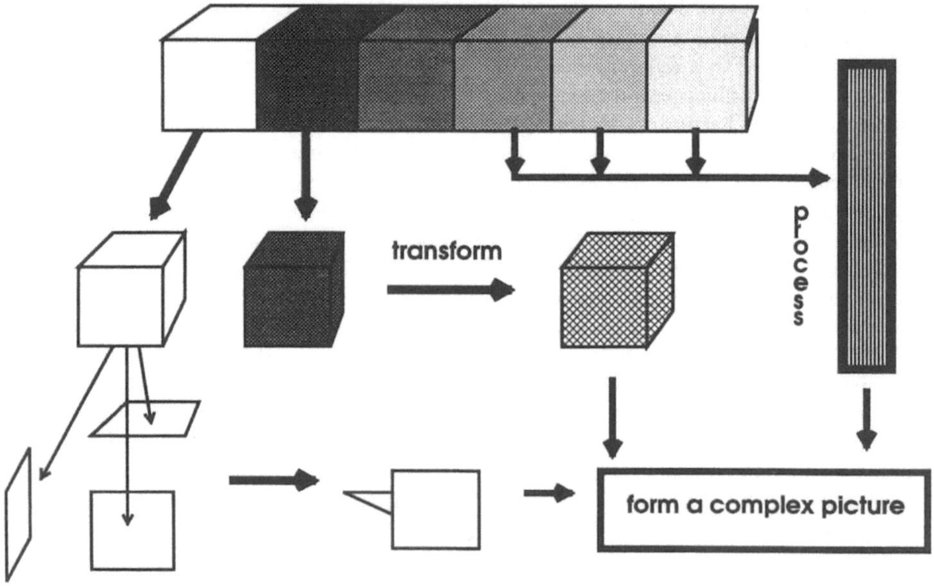

FIGURE 9.1. Processing of 3D fluid mechanic data

it will be necessary to extract the same slices from more than one block. These slices will be referred to as 2D-matrices from now on.

In many cases it is very convenient to deal with 2D-matrices instead of extracting the data directly from the original data set. In order to form more complex pictures, it will be necessary to combine graphics from more than one 2D-matrix into one graphic.

Another typical operation may transform data. For example calculating the Machnumber may need velocity data as well as density and temperature data. Such operation may be necessary on the block or slice level. This is indicated by the "transform" arrow, where one block of data is transformed into a block of different data. The new block still refers to the same original data set. This newly created 3D-matrix can be accessed in the same way as other 3D-matrices. Graphical elements may be created from the new data.

Sometimes very complex operation may need a larger number of 3D-matrices to be performed. The calculation of streamlines for example needs all coordinates plus all velocity components plus perhaps more data for additional effects. This may be a relatively CPU and I/O intensive job. The lines themselves will in general be combined with other graphical elements to form a complex picture.

For design purposes it will be necessary to get some information about the data size.

Figure 9.2 gives an overview about the data created for various flow problems. The calculation of the flow around the DFVLR-F5 wing with a finite volume Navier-Stokes code will be used as an example to demonstrate the capabilities of a graphic system later. This calculation of a transonic flow around a single wing mounted on a wall inside a windtunnel section may be considered to be a very small example. Still it creates almost 10 Mbytes of data. More complex problems like wing-body-configurations, especially when being in hypersonic flow, or complex configurations like fighter aircrafts easily create one or two orders of magnitude more of data. On the other hand it is clear that common 2D problems will not cause trouble from the data size point of view.

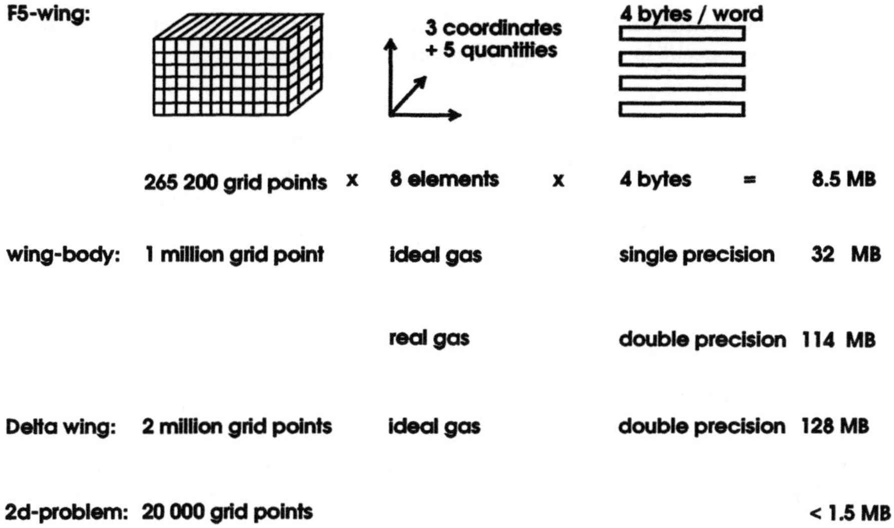

FIGURE 9.2. Data sizes for different flow problems

9.2.2 Modular software concept

About two years ago the institute decided to build a graphic system to run on it's Sun workstations. An analysis of the surrounding hardware and of the expected data almost automatically leads to a certain design concept for the software. Of course this concept may not necessarily be applicable for other installations. A different environment might lead to a different concept.

As most workstations do, Suns run a UNIX operating system. Software for UNIX very often follows the tool concept. A tool is, in some sense, a small program, which provides the user with a certain functionality. A tool exchanges data with other tools by writing files or reading files. Many tools seen together, form a complex system with large functionality. Often a series of tools are needed to give a certain result. They may be arranged to form a single command exchanging data using the UNIX piping mechanism. Tools increase the flexibility as they can be rearranged easily. The tool concept seems to be very powerful and is certainly one of the main reasons for the success of UNIX.

Flexibility

It turns out, that this tool concept can be efficiently applied to a graphic system as well. Graphic programs can be written as independent modules, thus increasing the flexibility of the system. Of course the modules need to exchange data with each other. This can be done using the UNIX file system at first. Some modules will be useful for more than one application.

Well-defined interfaces

Writing and reading files almost automatically defines a number of well defined interfaces. These modules are more easy to use in different applications.

Rapid prototyping

If only very few modules are created, they already form a working system. This is valuable for testing the concept in a very early phase of the realization.

Extensibility

New modules can be added to the system at any time. If the structure of the files used to transfer data from one module to another module is known, a new module may be put in between these modules. This does not make necessary any change of the existing modules. Extensions of the system can be done, even if the source code of the system is not available.

Maintenance

Modules can be kept small. This makes them easy to create and easy to maintain.

Mixing of programming languages

The file system used for data exchange represents a well defined interface for different programming languages. All programming languages provide a read and write mechanism. This allows using different languages for different applications. In the world of fluid mechanic engineers it will be a big advantage to write large parts of the system containing highly algorithmic code in Fortran 77 as this is the language, which engineers are used to. Fortran will ease maintenance and extension of the system, if this has to be done by the same people who work on the flow codes. On the other hand some things will be best done in a different language like C. The modular concept allows to write each module in the appropriate language without interference.

Nevertheless it may be necessary to mix languages within one module as well. Especially when trying to access the window system of the workstation. This should be restricted to the minimum number of cases.

Integrating modules from outside

Very often one will find a tool, which already performs a part of the job. The modular concept allows to integrate these foreign tools into a graphic system. These tools may come with the workstation vendor's software distribution, from a commercial software vendor or, very often in the UNIX world, from a public domain software source. There are various powerful tools for final layouting of graphics in the public domain. They have a high quality user interface and allow addition of schematic drawings or texts, editing of colors and more. There is often no need to reinvent the wheel.

9.2.3 Software layers

If it turns out, that device dependent parts of the software cannot be avoided, it is recommended to introduce appropriate interface layers in the software. Collecting all access to the window system within a limited number of routines for example, helps to switch to a new window system by simply mapping these routines to the new access functions.

Figure 9.3 suggests such software layers. The upper layer is represented by the window server. The window server takes control over all user action, i.e., all mouse and keyboard input. The window server manages the basic output control on the screen as well.

A program called SunTools is used as window manager. SunTools is a Sun proprietary software, which allows only local windows. For most graphic applications a network ex-

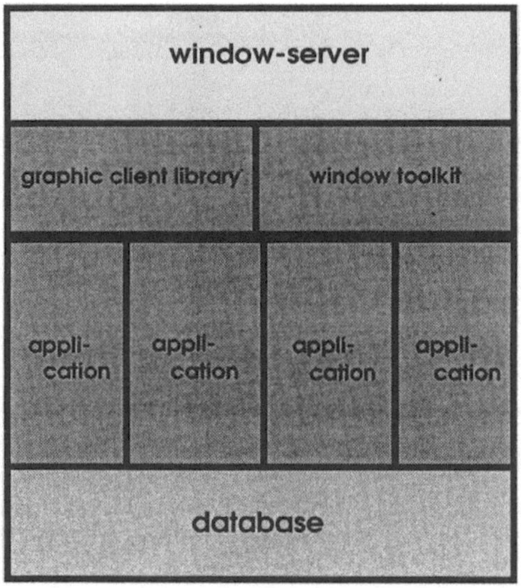

FIGURE 9.3. Software layers for a graphics system

tendable windowing system will not be necessary. The network will not be fast enough to transmit graphic information for these applications. But since the X11/NeWS window server seems to become a standard within the UNIX world, it will be advantageous to use it in the near future.

Actions of the window server are requested from the client (application) programs by calling functions. A set of such functions is called a window toolkit. For each window system there is at least one toolkit available. The SunView toolkit works directly with the SunTools window server

A second kind of output is generated on the screen. That is the graphics coming from graphic library. Sun supplies two different 3D graphic libraries at present, PHIGS and ACM Core. Both can access a special type of window, managed by the window server SunTools. At present HIGHEND is implemented on Sun's version of ACM-CORE. By making use of window servers, window toolkits and graphic libraries the graphic software can benefit from this modern technology. Unfortunately there is nor a window system neither a graphic library which is available on a large number of workstations from different vendors. So, the software designer should create an interface layer between his application on one side and the window toolkit and the graphic library on the other side, thus making later exchange possible. Inside the application layer different application modules may run in parallel. For data exchange they access files in the UNIX file system.

9.2.4 Overall architecture

The graphic system consists of a large number of modules. These modules work on a common database. The database is accessed automatically by the various modules. There is no need for the user to go into the details of the database.

All modules interact with the window system. They all together create a common desktop. The modular structure is hidden from the user. The naive user gets the impression

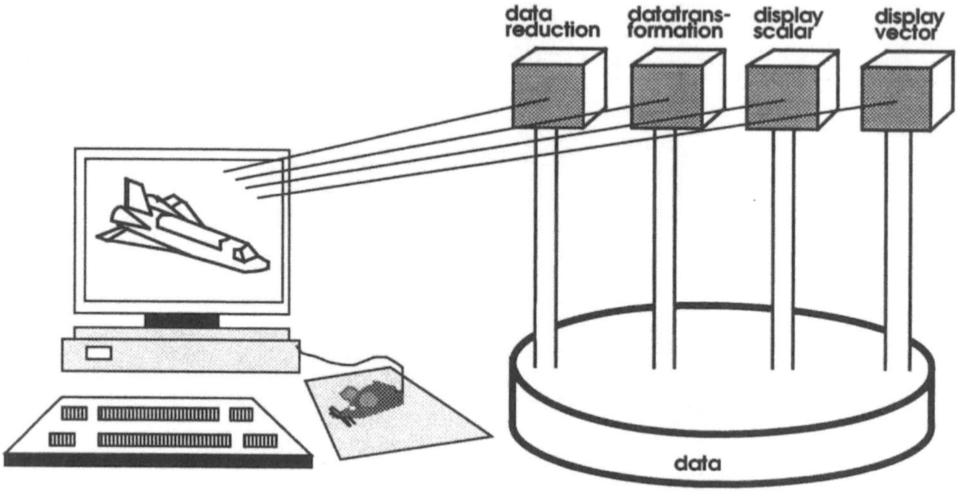

FIGURE 9.4. Overall architecture of the graphics system

of one solid system. The experienced user, who knows the details of the design, may make use of the flexibility of various modules.

Figure 9.4 illustrates the overall architecture. The modular concept allows the user to extend the system at any time. He simply writes his own module, which reads and writes from or to a data element in the data base. The new module can easily be integrated into the desktop. It will not be distinguished from the other modules. In this way new functionality may be added to the system even if no source code is available..

9.2.5 The desktop

The user interface created by a window system allows to rearrange work on the screen much similar to an office desktop. For this reason such comfortable user interfaces are called desktops. The desktop hides the internal complexity of the system from the user. Items on the desktop are managed by the window system. In order to create items on the desktop, functions from the window toolkit have to be called. If the same toolkit is used for the graphic modules as is used by other software running on the workstation, one can achieve a homogeneous appearance of all software. A modern phrase for this appearance is 'look' and 'feel'. Under the same name 'Look and Feel' Sun is now trying to present a catalogue of rules on how software sould be strucutred. When this system was designed the idea of 'Look and Feel' was not known to the public yet, neither was a toolkit available to implement it.

As the SunView toolkit is used by Sun supplied software as well as the graphic system a common look is guaranteed. All windows, buttons, menus, sliders and cycling knobs look the same and work in the same way for all modules. The use of the graphic system does not hardly require any typing, once the name of the original data set is made known to the system. Mouse input plays a major role. The window manager allows more than one graphic to be created at the same time on one screen. By this parallel work comparisons of different data can be easily done. On-line help texts are available for all the modules.

9.2.6 The database

A lot of different information will be stored in the database. First of all the scientific data set has to be kept for further processing. In addition to that there might be some descriptive data containing some information like additional parameters or the internal structure of the data.

After some processing has been done, this processed data will again be stored. For certain modules it makes sense to create temporary data in order to save valuable resources like main memory or computing time. This type of data will be kept as well. Finally some graphical objects will be created and displayed.

Some layout specifications are stored, to allow the user to create a series of pictures in the same style. Some modules will create a log file to save settings for later use. All this has to be stored.

In the present system all these data is stored in files in the UNIX file system. In order to free the user from the administration of this lot of files, the file names are created automatically when a file is written. The system knows the file names, when data is needed for read.

Starting from the name of the original data set, a tree like structure of files is created. In order to distinguish this data from the general data in the UNIX file system, which uses tree like structures as well, a different separating character (at present the "underscore" character) is used to separate hierachy levels.

Data exchange with the help of temporary files allows a very easy implementation of the modular concepts. More than one module can read the same data files. Files are very well defined interfaces, when all modules use the same access routines. Each file type represents a certain data structure. The file type and the type of the data structure can be checked for error detection. These interfaces may be used for adding more modules or as lower level entries to the system by the experienced user. If a file is missing, the system will automatically create it from the basic data set.

9.3 Capabilities of HIGHEND

9.3.1 User interface of the graphic system HIGHEND

In general the user interface is similar to all the other window tools in the Sun environment. Once a user knows how to work with mouse input on a Sun, he will easily find out how to work with the graphic system.

Plate 5 gives an example of some features of the user interface. When graphic elements are displayed, the user may rotate them to any viewing angle. During the rotation only the outer edges are displayed to keep the rotation fast enough for interactive work even on slow workstations. The object is rotated inside a graphic window by simply moving the mouse cursor inside this window

All parameters of the graphical layout may be controlled selecting on of the icons of the layout editor running on the left side of the screen. The icon menus give access to

- data-set name
- predesigned layout
- selected slices
- index range

- coordinate intervals
- minimum and maximum of the color table
- triggering
- solid surface shading
- hidden line
- wire frame
- background color
- surface color
- grid color
- text color
- light vector
- viewing angle
- size of graphic window

Changes are made visible inside the graphic window of the displaying module running on the upper right part of the screen. In the lower left corner of the screen a module for calculating various aerodynamic quantities displays its controlmenue. All modules run separately and may be placed at any position or size onto the screen. At present a german user interface language is implemented.

9.3.2 Calculating aerodynamic quantities

The original data-set in general contains only a minimum number of physical quantities, like three coordinates, three components of the velocity vector, the pressure and the temperature. The system does not imply any particular set or order of quantities. Nevertheless this is a commonly used combination. Other quantities of interest may be calculated from these, when needed. The system offers currently the following functions to be calculated:

- density
- pressure
- stagnation pressure
- pressure coefficient
- temperature
- stagnation temperature
- enthalpy
- inner energy
- Mach number

- entropy
- velocity components from momentum components
- momentum components from velocity components
- cross flow

These calculations are initiated with a button menu.

9.3.3 Thresholding

If a fast insight into 3D data is required, it is possible to select a series of slices inside this block of data. In order to avoid, that the slices in the front hide those behind them, a threshold value is set. Only those parts will be displayed, where the scalar value is inside the thresholding interval. For example regions where the scalar is very close to the freestream value may be skipped whereas regions showing significant changes will be selected. This mechanism lets the computer automatically select regions of interest.

Some quantities incorporate their own typical threshold levels. The Mach number M=1 is such a typical case. If this value is used a good impression of the supersonic region is created.

Plate 6 shows the Mach number inside the supersonic flow region above the DFVLR-F5 wing. The thresholding mechanism is a very powerful tool to isolate interesting phenomena inside a flow region. The wing is displayed in a solid surface Gouraud shaded mode to give an orientation in 3D space. A series of planes normal to the wing's surface is selected for Mach number display. The threshold level is set to 1. The color table is adjusted to give dark blue for M=1 and red for the highest Mach numbers inside the supersonic region. The selected Mach number interval goes from 1 to 1.3.

9.3.4 Streamlines

Calculating streamlines corresponds to the integration of a vector field. Doing this with HIGHEND involves three modules. The first module is used to select the streamline starting point inside the displayed domain. The second module actually calculates the streamlines and stores the vectors containing the streamline coordinates. If the domain is very large this job may be transferred to a faster machine. The third module allows to display the streamline, together with other graphical elements, such as surfaces or color coded scalar distributions. Plate 7 shows the vortical flow on top of a turbine blade.

9.4 References

[1] W Kordulla, D Schwamborn, and H Sobieczky. The DFVLR-F5 Wing Experiment. Technical Report AGARD CP437, 1988.

[2] Hans-Georg Pagendarm. A Typical Realization of a Graphic System for Fluid Dynamics. In *Computer Graphics and Flow Visualization in Computational Fluid Dynamics*, von Karman Institute for Fluid Dynamics, Lecture Notes 1989-07. Rhode-St-Genese, Belgium, 1989.

[3] H Zimmermann and D Schwamborn. Computation of Transonic Turbine Cascade Flow Using Navier-Stokes Equations. In *9th ISABE, 3–8 September, Athens*, 1989.

10 Supercomputing Visualization Systems for Scientific Data Analysis and Their Applications to Meteorology

Philip C. Chen

ABSTRACT
Two supercomputer-based scientific visualization systems implemented over a half-year period are augmented with graphics computers and peripherals. The comparison of these two systems shows that the animation efficiency is dependent upon individual computers: graphics capabilities and computation power; networking and file transfering efficiencies; memory management capacity; and software compatibility. A case is presented where applications of these systems by a meteorologist involves investigation of evolutions of weather systems. Based on his scientific knowledge, the meteorologist selected physically-related parameters for producing three- dimensional animations. The case study results show a composite animation with these related parameters. This animation reveals important weather system development mechanisms which could not have been realized by animations of individual parameters.

10.1 Introduction

Supercomputers have been used for model simulation studies for more than a decade. The treatment of model run results has been using computer graphics packages developed exclusively for supercomputer systems. Well-known computer graphics packages were developed in U.S. national laboratories, including the National Center of Atmospheric Research (NCAR), the Lawrence Livermore National Laboratory (LLNL), and the Los Alamos National Laboratory (LANL). Scientists running models in these laboratories have been using on-site graphics packages and hardware to generate graphics products for data analysis. In the past, graphics techniques used most frequently for data analysis include line-drawings, contours, and two-dimensional animation of meteorological parameters.

During the last decade, advanced graphics including three-dimensional visualization techniques emerged. However, three-dimensional data visualization was expensive and used by selected research activities only. Recently, advanced computer graphics hardware, software, and user interfaces have become less expensive and widely available. The U.S. national laboratories mentioned above as well as the National Center for Supercomputing Applications (NCSA) have developed advanced visualization software programs for scientific investigations [4]. Scientists began to use advanced visualization techniques for analyzing their data. For example, some meteorologists used supercomputer [3, 5, 1] with three-dimensional animation to study weather phenomena with several meteorological parameters displayed and depicted by color surfaces, volumes, and symbols. By viewing the animation of these parameters, a meteorologist infers the interactions among them.

In this paper, two supercomputer-based scientific data visualization systems, one using graphics workstations and the other using a mini-supercomputer, will be introduced. Meteorological parameters analyzed under such visualization systems will be presented. Efficiencies of these visualization systems will be discussed and evaluated.

10.2 Background information

10.2.1 Project description

An exploratory research project, started in August, 1989, has been using computer resources including: a supercomputer, a mini-supercomputer, and graphics computers. These project objectives are: conducting science investigations and evaluating visualization systems. The investigator of this project is a meteorologist, who is also a computer graphics expert.

10.2.2 Implementation phases of visualization systems

The Jet Propulsion Laboratory (JPL) Supercomputing Project installed a supercomputer Cray XMP/18 with two general-purpose Sun 4/280 workstations for Cray operations in June, 1989, and it became operational in August, 1989. The JPL supercomputer project did not provide graphics computers until January, 1990. Characterized by available graphics computer resources, this research project is divided into two phases as follows:

Phase 1: From August to December, 1989, using a Sun 4/280 and an IRIS 3030 graphics workstations borrowed from the JPL's laboratory facilities.

Phase 2: From January, 1990 to present, using a Stardent GS2000 mini-supercomputer and a Personal IRIS graphics workstation provided by the JPL supercomputing project's visualization laboratory.

More details of these two phases will be discussed later in Section 10.3.

10.2.3 Database and meteorological event

The database used in this research was provided by the European Centre for Medium-Range Weather Forecasts (ECMWF), U.K. The ECMWF database was generated by running the ECMWF numerical weather prediction model, using February 4, 1988, 1200 UTC global meteorological observation data as input. The model run provided 7 days of prediction data.

During this 7-day period, a weather event, i.e., a rapidly developing winter cyclone which originated near Newfoundland, travelled across the Atlantic Ocean, became intensified, remained stationary and produced heavy snow over the European Continent for several days. This winter cyclone was accurately predicted by the model. The ECMWF staff verified the prediction by using two-dimensional line graphics for comparing model output data and the observed data. The comparisons were showing good agreements between the observed and predicted temperatures; they were also showing good agreement between the observed and predicted sea- surface pressures.

Motivated by the successful numerical model prediction, the ECMWF staff prepared a fine-resolution database with an area covering the Atlantic Ocean and part of the European Continent for further studies. The ECMWF database contains 168 hours worth of forecast data including meteorological parameter data of geopotential, temperature, specific humidity, vertical velocity, vorticity, divergence, relative humidity, and horizontal u, v wind components. The parameter data were recorded in World Meteorological Organization (WMO) Gridded Binary (GRIB) format, which is an hourly data format.

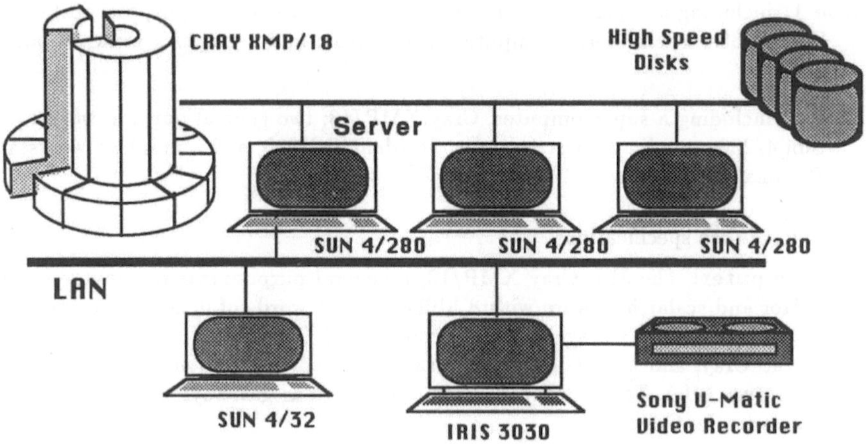

FIGURE 10.1. System Architecture for Phase 1

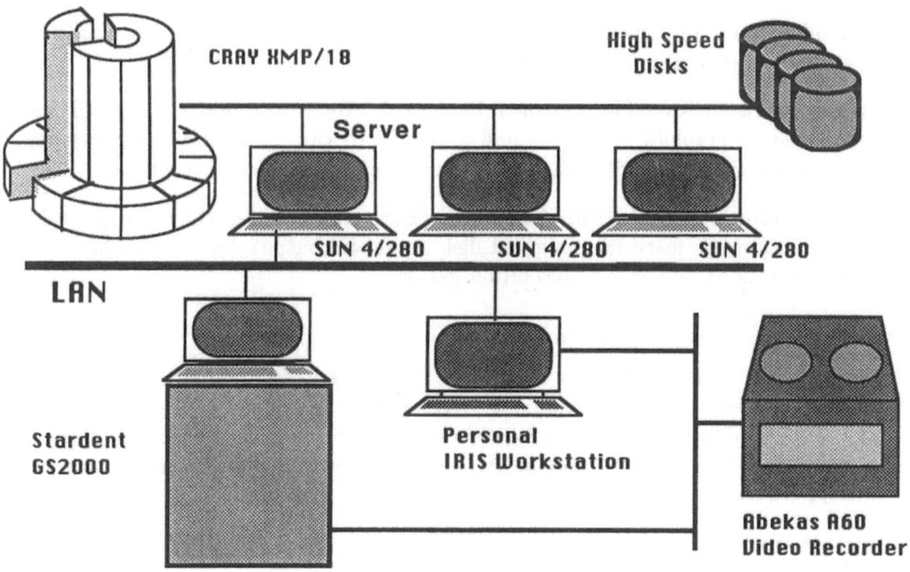

FIGURE 10.2. System Architecture for Phase 2

10.3 Computation and visualization systems

10.3.1 Hardware architecture

The hardware architectures, shown in Figures 10.1 and 10.2, are for:

Phase 1: including a supercomputer, Cray XMP/18; two general-purpose workstations, Sun 4/280s; a mini- supercomputer, a Sun 4/32; and a graphics workstation, IRIS 3030.

Phase 2: including a supercomputer, Cray XMP/18; two general-purpose workstations, Sun 4/280s; a mini- supercomputer, Stardent GS2000; and a graphics workstation, Personal IRIS.

10.3.2 Hardware specifications

Supercomputer: The JPL Cray XMP/18, a general purpose supercomputer featuring vector and scalar hardware with 8 Million 64-bit words of memory, can execute up to 200 MFLOPS. Storage devices include a 4.8 Giga Bytes disk attached directly to the Cray, and a 12 Giga Bytes disk connected to two Sun 4/280 via net work file server (nfs). These Sun processors also serve as gateways providing a network access to the Cray.

Mini-supercomputer: The Stardent GS2000 graphics computer, a mini-supercomputer designed exclusively for graphics applications featuring integrated multi-stream processor, vector floating point processor, and rendering processor hardware can execute up to a few tens MFLOPS. In graphics operations, this computer can render a few hundred thousand triangles per second, and it can handle near-realtime interactive graphics instructions.

Graphics workstation: The Personal IRIS workstation, a graphics computer featuring custom graphics processors and a fast CPU, can render moderate numbers of polygons and can handle near-realtime interactive graphics instructions.

General purpose workstation: The Sun 4 series workstations, which are general purpose computers featuring tape and disk input and output devices, can handle slow access data as well as low- resolution graphics.

10.3.3 Operating system and windows

Supercomputer: The JPL Cray runs the UNICOS, a Cray-proprietary operating system based on the AT&T System V UNIX operating system. UNICOS incorporates several modifications for supercomputers, including enhanced security features, a redesigned file system to support large files, improved resource control features and support for TCP/IP access.

Mini-supercomputer: The Stardent GS2000 runs the AT&T and Berkeley System Distribution (BSD) UNIX operating systems, and runs applications on X-Windows.

Workstations: Sun and IRIS workstations run the AT&T and Berkeley System Distribution (BSD) UNIX operating systems and applications on vendor-supplied window systems, but could run on X-Windows as well.

10.3.4 Software specifications

Supercomputer: The JPL Cray software includes compilers: Fortran, C and Cray Assembler; debugging tool: CDBX; mathematical library: IMSL; Cray libraries: MATHLIB and SCILIB; graphics packages: MOVIE.BYU, NCAR Graphics, and Cray graphics system – OASIS. The OASIS (Our Animation-Simulation Interactive System) was provided by Cray Research, Inc. This software system has programs including: The Interactive Modeler (TIM) – modeling and scene control; the Clockwork – digital video frame generation using techniques of raytracing and volume transparency; and the Display – image display, animation and recording. This system was implemented on Sun and IRIS workstations as well as on Cray machines, and it was used to do data visualization and production of animations during phase 1 of this project.

Mini-supercomputer: The Stardent GS2000 software includes compilers: Fortran and C; debugging tool: dbx; mathematical library; graphics libraries: XDI – X-window based PHIGS+ graphics routines, and Application Visualization System (AVS) – a collection of high-level interactive graphics programming tools. The AVS has capabilities including modeling, rendering, frame generation, display, and animation; the AVS is built upon the X-Window system, which provides highly interactive mouse-driven user-interfaces.

Graphics workstation: The IRIS workstation software includes compilers: Fortran and C; debugging tool: dbx; mathematical library; graphics libraries: Cray graphics system – OASIS available on an IRIS 3030, and Graphics Library (GL) – a collection of graphics routines provided by Silicon Graphics, Inc.

General purpose workstation: The Sun workstation software includes compilers: Fortran and C; debugging tool: dbx; mathematical library; graphics libraries: Cray Graphics – OASIS available on selected workstations, and Sun provided graphics routines.

10.3.5 Networking and computer resources utilization

All computers used for this research have been connected to a Local Area Network (LAN), so that data could be transferred to/from any computer and graphics software programs could be executed on any computer.

In practice, however, using essentially a distributed approach, each type of computer was used for special purposes: in phase 1, the Cray supercomputer for data preparation and video frame generation, the Sun workstation for animation preview and image file storage, and the IRIS workstation with SONY U-Matic deck for video recording; in phase 2, the Cray was used for data preparation and the Stardent GS2000 and Personal IRIS for visualization/animation, and Abekas A60 video system for recording.

10.4 Parameter selection, derivation and data preparation

10.4.1 Parameter selection and derivation

While in computer graphics character animations a character (or an actor) is well-defined and controllable, in scientific visualization animations an entity (object) which may be of interest to a scientist is usually discovered a posteriori – after an animation has been produced [1]. This after-the-fact determination of an animated object makes parameter selection difficult. However, scientists may choose parameters based on experiences acquired from previous work, scientific literature, research papers and textbooks, etc., and use scientific visualization techniques to produce an animation. Likewise, experienced meteorologists often choose parameters including winds, temperature, trajectories, vorticity, divergence, clouds, etc., and use visualization techniques including contouring, volume rendering, transparency to produce animations.

This research started out by animating ordinary numerical model input/output parameters such as winds, temperature and geopotential. However, it was soon realized that animating these common parameters would only lead to the traditional understanding of evolution of weather systems. Therefore, as an experiment and departing from traditional meteorological practice, parameters selected for animations included: kinetic energy and potential temperature, which are often neglected by meteorologists; and water vapor specific humidity. Reasons for selecting these parameters will be elaborated on later in this section.

The ECMWF database does not provide kinetic energy and potential temperature data because these are not model output data. However, these parameter data can be derived from available model output data. The kinetic energy is computed by using horizontal wind components:

$$K = 1/2 * (u2 + v2)$$

where K is the horizontal kinetic energy, u the easterly wind component and v the northerly wind component. The potential temperature is computed by using pressure and temperature:

$$P = T(p/P_0)k$$

where P is the potential temperature, T the atmospheric temperature and p the atmospheric pressure. p_0 is the reference atmospheric pressure and k is a meteorological constant, 0.286. The water vapor specific humidity data need not be computed, as they are available in the ECMWF database.

In this research, the selection of parameters is based on energy conservation law and energy transfer between different energy forms. By selecting kinetic energy, one can perceive how the kinetic energy was transported horizontally and vertically, and how it would relate to cyclone formation. By selecting potential temperature, one can see how thermodynamic quantity, entropy or heat content, is related to the weather system. By selecting specific humidity, one can visualize the outline and structure of a weather system.

10.4.2 Data preparation

The ECMWF database size is characterized as follows: each parameter, which has data for 91x47 horizontal grids and 14 pressure levels, has about 60K data values for each hour. For 7 days, i.e., 168 hours, there are about 10 Million data values. Provided that each data value is stored with 32 bits (4 bytes), the size of a 168-hour database is about 40 Mbytes. For example, in a database containing 9 parameters, there are about 360 Mbytes, a formidable data size for a computer to accommodate.

The ECMWF database uses WMO GRIB data format which is great for data archiving, but not good for retrieving a time varying parameter. Thus, the database was further processed by Cray supercomputers, which are equipped with large storage devices, until each meteorological parameter had its own dataset. With these individual meteorological parameter datasets available, further data processing including computations of new parameters and generating parameter visualization image became quite easy. It was not necessary to go through voluminous sequential, hourly, multiple meteorological data to retrieve a particular meteorological parameter or parameters; instead one could obtain data from smaller processed datasets.

10.5 Animation production procedures used in phase 1

Animation production in phase 1 has been done basically on the JPL Cray, where digital frames were generated. These frames were transported via network to workstations, previewed on workstations, and recorded on a workstation-video recorder device. The detailed procedures are elaborated as follows:

10.5.1 Contour value, viewing perspective, and lighting

In this research, an image was created regarding contour values, viewing perspective, and lighting. The object-image creation steps are:

- Creating a contoured data file: A three-dimensional parameter data, extracted from a parameter dataset, is pre- processed with a contour value to a data file with a format acceptable to the Clockwork.
- Creating a scene file: A scene file defining viewing perspective and lighting condition is prepared.
- Creating an image file: The scene file and data file are input to the Clockwork. A graphic image file containing three-dimensional rendered objects is created by the Clockwork.

The object-image creation begins by following the steps and running the Clockwork on the JPL Cray with a set of trial contour value, a viewing perspective and lighting conditions to produce an image file. Then, using the Display, this image file is downloaded to Sun or IRIS workstations for viewing. The object-image creation and viewing process is repeated until a satisfactory image is obtained.

Once determined, the contouring values, perspective, and lighting will be applied to all images to be generated for animation production.

10.5.2 Animation preview

Since the Sun and IRIS workstations used for this research did not provide fast playback capability, animation preview was accomplished by viewing images at a low frame rate and a low image resolution. The animation preview is elaborated as follows:

A low spatial and temporal resolution animation with sequential video image frames produced by the JPL Cray were downloaded to a graphics workstation. Frames generated in this animation have low pixel resolution, typically 128x128 pixels, and were produced by skipping a few records. Using a prescribed script file as an input to the Display program, these frames were automatically displayed on a workstation. The motion of objects displayed with a low raster rate, about a frame per second, could be perceived by paying special attention to object displacements. If the motion proved satisfactory, one proceeded to the next procedure. Otherwise, it was back to the previous procedure for determining contours, viewing perspective, and lighting.

10.5.3 Video production

In this procedure, the animation sequence of a parameter or parameters was produced on the JPL Cray with full spatial resolution (512x512 pixels) and fine temporal (hourly) images. This animation sequence consists of rendered video images generated by raytracing with volume transparency. Generated video images in tri-color (red, green, and blue) format were transported via network to workstations and recorded onto a video tape.

10.6 Animation production procedures used in phase 2

The high resolution animation production following procedures used in phase 1 has been described in Section 10.5. In addition to the high resolution animation production, using Stardent GS2000 for low resolution animation the procedures have been modified. The detailed procedures are elaborated as follows:

10.6.1 Contour value, viewing perspective, and lighting

In this research, a digital image was created regarding contour values, viewing perspective, and lighting. The object-image creation steps are:

- Creating a contoured geometry file: A three-dimensional parameter data file is preprocessed by a Cray contour generating program to create a geometry data file with a format acceptable to the AVS. This geometry data file, consisting of a polygon vertex list of contoured objects, is then downloaded to the Stardent GS2000.
- Creating a scene file: A scene file defining viewing perspective and lighting condition is prepared either by a command file or interactively.
- Creating an image: The scene and geometry data files are input to the AVS to generate a graphic image with three- dimensional, rendered, contoured objects.
- Saving scene parameters: The AVS tools are used to experiment lighting, viewing perspective, camera angle, objects positioning, etc. The scene parameters will be saved to a scene file issued by an user.

The above image-object creation steps are repeated until a satisfactory image is obtained. Once determined, the contouring values and scene parameters, including perspective and lighting will be applied to all images to be generated for animation production.

10.6.2 Animation preview

The Stardent GS2000 provides fast playback capability. Animation preview was accomplished by viewing contoured objects in reduced frame rate and reduced parameter data, typically at grid resolutions of 46 x 23 x 14. The animation preview is elaborated as follows:

A sequence of low-spatial resolution geometry files created by the JPL Cray was downloaded to the Stardent GS2000. An animation script file was created by an AVS preprocessing program using these geometry files as an input. By using this script file, an animation can be displayed automatically on the Stardent GS2000. If the motion proved satisfactory, one proceeded to the next procedure. Otherwise, it was back to the previous procedure for determining contours, viewing perspective, and lighting.

10.6.3 Video production

At the present time, no video has been produced using Stardent GS2000 or Personal IRIS computers. These two computers will be linked to an Abekas video recording system, where video images created by these two computers will be recorded frame-by-frame.

10.7 Data analysis results

10.7.1 Phase 1: parameter animation results

Parameter animation results including kinetic energy, water vapor specific humidity and potential temperature will be shown at the conference video presentation. Features of animation results are shown in Plates 8 to 10. The base map, showing the geographical area coverage of the Atlantic Ocean and Western Europe, is shown in cyan. The outline of Greenland is visible in Plates 8 to 10.

Plates 8 to 10 show composite fields of kinetic energy and water vapor specific humidity at predicted hours of 113, 125 and 135. These hours correspond to times on February 9 to 10, 1988 when the cyclone in question was in the fullest developmental stage. For the kinetic energy field: cloud-like white features, with some flaunting red enclosures, are located near the top of troposphere and stratosphere and contain high energies. Two energy categories, from 1,000 to 2,000 and from 2,000 to 3,600 m2/sec, are colored in white and red. For the water vapor specific humidity field: terrain-like, yellow and purple features near the surface of earth contain water vapor. Two water vapor specific humidity ranges, from 3 to 6 g/kg and from 12 to 22 g/kg, are colored in yellow and purple.

Plate 11 is a snapshot of potential temperature at the 120th hour when the cyclone became nearly stationary. The potential temperatures, which range from 295 to 305K, are colored in white. Results obtained from repeatedly viewing the video tape reveal important cyclonic structural features relating to development mechanisms. The results that can be concluded are listed as follows:

- Kinetic energy field shows tropospheric jet-like structure with near concentric energy distributions. Examples of such structures are shown in plates 8 to 10, colored in white and red. Animation of kinetic energy shows that the jet structure changes shape constantly, as is shown in plates 8 to 10. The wedge-like configuration is generally associated with a cyclone on the ground.

- Water vapor specific humidity field shows structures resembling terrain: a summit indicates a cyclonic storm center, and a ridge/trough indicates a frontal zone in three dimensions. Examples of such a storm structure are shown in plates 8 to 10, colored in yellow and purple. Animation of water vapor specific humidity shows that iso-surfaces are driven by winds, and the motion of surface is three- dimensional, with evidence of vertical oscillation relating to gravity waves.

- Potential temperature field and water vapor specific humidity field show near-circular structures associated with a cyclonic storm at the ground. Examples of such a structure are shown in plates 10 and 11.

- Animation of potential temperature shows that the near- circular structure at one time became nearly stationary, and motion was confined to a downward direction. This is evidence of an air subsidence during a later stage of cyclonic life cycle.

- As realized in plates 8 to 10, composite animation of kinetic energy and water vapor specific humidity shows that the two fields are closely related. The evidence of an atmospheric jet feeding energy to a cyclone near the ground for its further development can be seen from the composite animation with a white-colored, funnel-like feature, as shown in plates 9 and 10.

10.7.2 Phase 2: work in progress

Meteorological data visualization research work in progress includes:

- Solving the problem of three-dimensional spatial ambiguity by designing and implementing visual aids such as three- dimensional cursor, digital clock, high-lighter, rosettes, compass, multiple windows of cross-sectional views, etc;
- Investigating accuracies of graphics algorithms including contouring, volume rendering, and trajectory;
- Using interactive high-speed graphics workstation for visualization more parameters and more cases;
- Making real-time data value retrieval available for a meteorologist during an interactive visualization session;
- Investigating physical laws and select parameters for animation.

10.8 Visualization system evaluations

Evaluations for two-phase systems are highlighted as follows:

Phase 1 used the JPL Cray XMP/18 for major computations including object rendering and frame generation, and workstations for image previewing and video recording. The efficiency of this distributed visualization system relies heavily on efficiencies of the Cray software and the recording device. The Clockwork generates a high-resolution, ray-traced image at a rate of 20 minutes/frame. Video recording was done on the IRIS 3030 connected to U-Matic Deck, with a video recording speed at 2 minutes/frame. A 168-hour single parameter animation will take at least 3 days to complete. With man-power constraint and other operational overhead, an animation per week is about the average production rate for this system. A more efficient run of the Clockwork can be achieved by using Cray YMP or Cray2. The run time can be reduced to approximately a few minutes/frame. With this rendering speed, the animation may take less than 3 days to be produced.

Phase 2 used the JPL Cray for generating object geometry files and the Stardent GS2000 for rendering and animation. This distributed visualization system put less burden on the Cray and more on the Stardent GS2000. The efficiency of this distributed system depends heavily on the efficiency of the Stardent GS2000 hardware and software. The Stardent GS2000 performance is limited by available memory and AVS capabilities. The Stardent GS2000 memory is 64 Mbytes, and the AVS has to load all frames to be animated into the memory. This small memory and AVS feature seriously impair the animation capability. The original dataset will have to be reduced in space and time in order to be animated with a reasonable animation speed such as 8 frames/sec. Realizing these shortcomings, the Stardent GS2000 could be used very effectively by applying its highly interactive tools that control lighting, object orientation and perspective for investigation of an object structure, such as a weather system.

10.9 Conclusions

On visualization systems:

- Supercomputing visualization system application: For data continuity and expedience, it makes sense to use the supercomputer-based, distributed visualization system. The advantage of using this visualization system has been discussed elsewhere [2]; it chiefly shortens the video production time to the extent that phase 1 was accomplished within four months.

- Individual vs collective efficiency: The efficiency of a distributed visualization system cannot be enhanced by merely connecting together efficient machines, as in the phase 2 study.

- Mini-supercomputers are good for interactive low spatial and temporal resolution animations. High resolution animations still rely on supercomputing.

- The animation efficiency may be raised: by using better networking technology such as FDDI, ultra-net, or other ultra high-speed connections; by sharing memory among computers with memory management techniques; and by designing compatible software for different machines, so that concurrent executions are possible.

- Future perspective: Designers of supercomputers, mini- supercomputers, and super-graphics workstations certainly will pay attention to these animation efficiency enhancement issues targeting real-time visualization systems.

On scientific visualization applications – meteorological analysis:

- Data visualization: Composite visualization should be based on physical laws involving key parameters.

- Future perspective: More meteorological analyses will be done with three-dimensional animations of parameters.

Acknowledgements:

The author wishes to thank the ECMWF which provided me with initial research time and a database. The author also wishes to thank Cray Research, Inc. for providing staff and material supports. Thanks are extended to Jet Propulsion Laboratory, Supercomputing Project gave the author free access to Cray X-MP/18.

10.10 References

[1] P C Chen. Computer Graphics Animation Techniques and Production Procedures for Scientific Visualization. In *Proceedings of the Workshop on Graphics in Meteorology, November 31 to December 5*, pages 56–65, European Centre for Medium-Range Weather Forecasts, UK, 1988.

[2] P C Chen. Computer Graphics Systems. In *Proceedings of the Workshop on Graphics in Meteorology, November 31 to December 5*, pages 111–132, European Centre for Medium-Range Weather Forecasts, UK, 1988.

[3] R Grotjahn and R M Chervin. Animated Graphics in Meteorological Research and Presentations. *Bull. Amer. Meteo. Soc.*, 65:1201–1208, 1984.

[4] L J Rosenblum. Scientific Visualization at Research Laboratories. *IEEE, Computer*, 22(8):68–70, August 1989.

[5] R Wilhelmson, L Wicker, H Brooks, and C Shaw. The Display of Modeled Storms. In *Fifth Intl. Conf., Interactive Information and Processing Systems for Meteorology, Oceanography, and Hydrology, January 30 to February 3*, 1989.

Part IV

Rendering techniques

11 Rendering Lines on Curved Surfaces

Jarke J. van Wijk

ABSTRACT
The visualization of functions over surfaces is a classical application of scientific visualization. Several techniques, such as grid-lines, contours, and smooth shading, can be employed, each with their own advantages. The result of the combined use of those techniques is often disappointing because the lines, drawn in image space, are unrelated to the surface, rendered in object space. A better result can be achieved if the lines are modelled in three dimensions. A model for lines as semi-transparent tape stuck on the surface is presented. Its integration in rendering systems is discussed, and an implementation and results are presented.

11.1 Introduction

The visualization of two-dimensional functions $f(x, y)$ is a classical application of computer graphics. As an example, already the second figure in the well-known handbook by Foley and van Dam[6] shows a three-dimensional plot of a function of two variables as a familiar example of computer graphics. The reasons for the popularity of this kind of graphs are obvious. In many disciplines, and especially engineering, functions of two variables are routinely used, and graphics are much more easy to grasp than alternative means such as tables.

The standard representation of a function of two variables is as a 3D surface, with the height proportional to the value of the function. Plate 12 shows four different visualizations of such a surface. The function used here was defined by a B-spline approximation of 8 *times* 16 randomly chosen values. At the boundaries of the rectangular domain the value of the function was set to zero. Such an artificial data-set is a good test for visualization techniques: the eye is not guided by a simple structure in the data.

Plate 12a shows the most popular representation of a 2D function: a net of lines with constant u and v values, or iso-parameter lines. As a result, the course of the function for constant parameter values can easily be followed. A disadvantage is that the height of the surface cannot be assessed accurately. A solution is to change the view to a lower viewpoint and scale the height of the surface, but this will lead to peaks that obscure the surface behind.

The contour lines or iso-function lines, shown in plate 12b, provide a better cue for the height. Local minima and maxima can be found easily. For the location of global extrema continuous tone techniques are more convenient. Plate 12c shows the use of eight grey-shades to indicate the height.

The previous techniques share that gradient information, i.e. on the steepness and direction of the surface, can only be derived indirectly, namely by looking at the distance between adjacent lines. An alternative approach is shown in plate 12d. Here a light-source is defined and the diffuse and specular reflection of light on the surface are modelled. The realistic appearance of such a shaded surface makes it easy to judge its shape: the eye has had a thorough training in the interpretation of such images. However, the advantages of the discrete methods based on lines are lost: the value and the position of extrema cannot be assessed quantitatively, and the course along iso-parameter lines cannot be followed.

This suggests to use several techniques simultaneously. However, the result of the superposition of a line drawing over a shaded image is disappointing. Especially when grid lines and contours are shown simultaneously visual chaos results: the smooth shaded image is cut into pieces, and the lines are hard to follow. The appeal of plate 12d suggests another approach and leads to the theme of this paper. Visual realism gives an image that is easy to grasp, so why not model lines as real, physical objects? In other words, when lines are drawn on the surface (in object space) a better result can be achieved than when lines are drawn in image space.

In section 11.2 the modelling of lines drawn in 3D space is discussed. In section 11.3 the integration of the model with rendering algorithms is discussed, while in section 11.4 an implementation and results are presented. Finally, in section 11.5 conclusions are drawn and suggestions are given for further research.

11.2 Modelling lines in three dimensions

Two techniques can be used for the modelling of lines in 3D space. First, lines can be considered as separate geometric elements. In a preprocessing phase the surface and its associated functions are analyzed, and geometric elements (polygons, bicubic patches) are generated that represent the lines.

For iso-parameter lines this calculation is straightforward, but for contours the associated geometric calculations are less trivial. In both cases, however, the selection of the distance or offset of the elements from the surface is prone to errors. For many surfaces a constant offset from the surface is hard to realize: the offset surface of an ellipsoid for instance is in general not an ellipsoid. If the offset is too small, the line is likely to penetrate the surface at some places and become invisible. If a large offset is chosen, there is a risk that the gap between the surface and the line will be visible, particularly with close-up views in animations.

The second, better approach is to use texture mapping techniques, i.e., to integrate the rendering of the lines in the shading of the surface. This requires that for each point on the surface two questions can be answered: does this point belong to a line, and if so, what is its colour?

To answer the first, geometric question, a surface is defined as a mapping from two-dimensional R^2 space with coordinates u and v to three dimensional euclidean space E^3:

$$\mathbf{x}(u,v) = [x(u,v), y(u,v), z(u,v)]$$

The scalar function $f(u,v)$ to be visualized is defined over the same two-dimensional space R^2, so each pair of coordinates $[u,v]$ defines a point at the surface as well as one or more values of attributes of that point. Special cases of $f(u,v)$ are $f(u,v) = u$ (iso-parameter lines), and $f(u,v) = z(u,v)$ (height lines). This definition is more general than the simple definition $z = f(x,y)$, because it also includes scalar functions defined over arbitrary surfaces. Examples of applications are the temperature, pressure, and bending stress over mechanical parts, such as beams, ship hulls, and wings.

The lines on the surface can be described by:

$$f(u,v) = C, \text{with } C = kD, \quad k = \ldots, -2, -1, 0, 1, 2, \ldots$$

where D gives the increment between values for which lines have to be drawn. Such a line is infinitely thin, which is not practical for visualization. As a first approximation a margin *epsilon* in terms of $f(u,v)$ can be used. The Boolean function $L(u,v)$ that

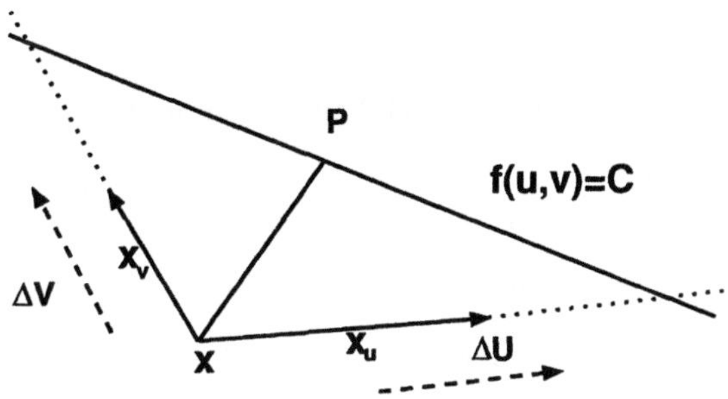

FIGURE 11.1. Calculation of distance from line

indicates whether point $\mathbf{x}(u, v)$ belongs to a line $f(u, v) = C$, is thus given by:

$$L(u, v) = |f(u, v) - C| < \epsilon/2$$

With this definition, the line width in 3D space is not constant, but dependent on the magnitude of the gradient of the function. This is acceptable if the function has a geometrical meaning, such as height. The surface is shown as if it was cut out from layered material, such as plywood: a physical analogue that allows for an easy interpretation. If the function denotes for instance pressure, shown over the surface of an aeroplane, the varying width is much harder to interpret.
A constant line width in 3D space is a better solution. This requires the calculation of the distance $d(u, v)$ of a point in 3D space to the centre of a line. The exact calculation for general surfaces and functions is not simple. However, provided that the lines are thin, compared to the curvature of the surface and the function, this calculation can be simplified via the use of linear approximations for $f(u, v)$ and $\mathbf{x}(u, v)$. For the calculation of d for a point with coordinates $[u_0, v_0]$ the function $f(u, v)$ is approximated by:

$$f(u, v) \approx f(u_0, v_0) + \frac{\partial f(u_0, v_0)}{\partial u}(u - u_0) + \frac{\partial f(u_0, v_0)}{\partial v}(v - v_0),$$

or, simplifying notation,

$$f(u, v) \approx f + f_u \Delta u + f_v \Delta v$$

Similarly, the surface $\mathbf{x}(u, v)$ is approximated by:

$$\mathbf{x}(u, v) \approx \mathbf{x} + \mathbf{x}_u \Delta u + \mathbf{x}_v \Delta v,$$

To calculate $d(u_0, v_0)$ a point $\mathbf{p}$ on the surface $\mathbf{x}(u, v)$ has to be found for which two conditions hold (see figure 11.1).
First, the point has to lie on the line; second, the distance to $\mathbf{x}$ has to be minimal. The point $\mathbf{p}$ lies on $\mathbf{x}(u, v)$, and can therefore be defined by:

$$\mathbf{p} = \mathbf{x} + \alpha \mathbf{x}_u + \beta \mathbf{x}_v$$

The first condition gives

$$f + \alpha f_u + \beta f_v = C \tag{11.1}$$

The second condition is satisfied if the vector from $\mathbf{x}$ to $\mathbf{p}$ is perpendicular to a vector along the line, or

$$(\mathbf{p} - \mathbf{x}) \cdot (f_v \mathbf{x}_u - f_u \mathbf{x}_v) = 0$$

or

$$(\alpha \mathbf{x}_u + \beta \mathbf{x}_v) \cdot (f_v \mathbf{x}_u - f_u \mathbf{x}_v) = 0 \tag{11.2}$$

The combination of equations 11.1 and 11.2 gives a linear system in two unknowns:

$$\begin{bmatrix} f_u & f_v \\ f_v \mathbf{x}_u \cdot \mathbf{x}_u - f_u \mathbf{x}_u \cdot \mathbf{x}_v & f_v \mathbf{x}_u \cdot \mathbf{x}_v - f_u \mathbf{x}_v \cdot \mathbf{x}_v \end{bmatrix} \begin{bmatrix} \alpha \\ \beta \end{bmatrix} = \begin{bmatrix} C - f \\ 0 \end{bmatrix}$$

the solution of which gives α and β. This gives for the distance of $\mathbf{x}$ to the line:

$$d(u_0, v_0) = |\alpha \mathbf{x}_u + \beta \mathbf{x}_v|,$$

which can be used for a new definition of $L(u, v)$

$$L(u, v) = d(u, v) < w/2$$

where w denotes the width of the line in object space.

The calculation of $d(u, v)$ can be simplified if a priori assumptions may be made about $\mathbf{x}(u, v)$ or $f(u, v)$. For instance, if $\mathbf{x}_u$ and $\mathbf{x}_v$ are orthogonal,

$$d(u_0, v_0) = \left| \frac{C - f}{f_u^2(\mathbf{x}_v \cdot \mathbf{x}_v) + f_v^2(\mathbf{x}_u \cdot \mathbf{x}_u)} (f_u(\mathbf{x}_v \cdot \mathbf{x}_v)\mathbf{x}_u + f_v(\mathbf{x}_u \cdot \mathbf{x}_u)\mathbf{x}_v) \right|$$

If $\mathbf{x}_u$ and $\mathbf{x}_v$ may be assumed to be (approximately) orthonormal, this reduces further to

$$d(u_0, v_0) = \frac{C - f}{\sqrt{f_u^2 + f_v^2}}$$

With these definitions it can be assessed whether a point belongs to a line or not. The next question is how the appearance of the surface is modified in order to show the line. The colour of the surface, as shown on the screen, is dependent on a set of parameters, such as coefficients for diffuse and specular reflection, the roughness of the material, the direction of the normal on the surface, and the position and the intensity of the light sources. Via a shading model[6] the red, green and blue values to be displayed are calculated.

With texture mapping techniques the appearance of the surface is modified via the definition of one or more of the shading parameters as a function of the position on the surface. The first, and still most popular application of texture mapping[2] involves the coefficients for diffuse reflection: an image is painted on the surface. Blinn has shown that via perturbation of the normals wrinkled surfaces can be modelled[1]. In[3] the mapping concept has been generalized further: it shows that mapping can be applied to all parameters, even the geometry of the surface itself, and that this requires a flexible environment for experiments, i.e., shade trees.

Similar techniques can be used for rendering lines. The special requirements for this application are that the lines must be distinguishable from the surface, they must be subtle, and realistic. The first requirement is obvious, but not trivial. For instance, if a dark line is used, the contrast will be too low at those parts of the surface where only ambient light is present. The modification of the roughness and specular reflection of the surface, f.i. rough lines over a polished surface, gives an aesthetically very pleasing effect, because sometimes the lines are brighter than the surface, and sometimes the reverse occurs. At the transition areas, however, the contrast will be too low.

Solid, bright red lines will be visible everywhere (provided that bright red is not used for the surface), but such lines will be too dominant, and distract the eye from the shape of the surface. In[10] the use of grey grid-lines is recommended instead of black lines. A subtle solution is hence required. However, too subtle lines are in conflict with the first requirement, so preferably the contrast between the surface and the line should be tunable for the particular image or application in hands.

The lines are easier to grasp if they simulate a real-world, physical model. For instance, the modification of the saturation or the green-component of the colour of the surface has no straightforward physical counterpart, and therefore requires an explanation, and some training of the viewer.

A solution that satisfies the preceding requirements is to model the lines as semi-transparent, coloured tapes. The amount of transparency controls the contrast between the line and the surface, while the tape itself can be coloured to provide contrast at dark areas. An elaborate model should incorporate the angle at which the line is seen, but a simple model suffices in practice:

$$k_{result} = k_{surface}t + k_{line}(1 - t),$$

where t denotes the transparency coefficient, and k denotes the diffuse reflection factor of a colour component for the result, the surface, and the line.

11.3 Integration with rendering algorithms

The preceding model can be implemented in several ways. Different trade-offs between flexibility, speed, and memory usage can be made. Further, the implementation depends on the hidden surface algorithm used. Two main approaches can be distinguished: precalculation of the texture map, and procedural mapping during the rendering.

The first approach involves the a priori calculation of a discrete texture map in uv space, which is mapped on the surface afterwards. A typical resolution of such a texture map is 256 × 256 pixels. This approach has several advantages. First, calculations can be done in the natural u, v order of the surface. Hence, incremental methods can be used to speed up the calculations. If the size of the steps in u and v direction is small and the same, a further acceleration can be achieved. The partial derivatives do not have to be calculated explicitly, but instead the values of the increments for f and $\mathbf{x}$ can be used for most of the calculation of $L(u, v)$. A further, generally applicable improvement is to calculate the square of the distance d instead of d itself, which saves the extraction of a square root.

The second advantage is that calculations have to be done only once for a particular surface and an overlaid function: if the projection is changed, the lines do not have to be calculated anew. For animation of dynamic surfaces and functions, however, this advantage does not hold.

The third advantage is that standard techniques for anti-aliasing can be used. Prefiltering techniques such as pyramidal parametrics[12] or summed-area tables[5] can be used during the rendering to remove aliasing effects, such as jagged edges and broken lines.

Probably the most important advantage of precalculation is that the implementation of 3D lines does not require modification of the rendering software, provided that it has facilities for texture mapping.

The other approach is to apply the model described here as a part of the shading of the surface during the rendering. The complexity of this step is dependent on the hidden surface algorithm used. The worst case is ray-tracing[11]. Here the surfaces are sampled without coherence in object space, hence for each point nearly all calculations have to be done anew.

The rendering algorithm used for Reyes[8], the rendering system of Pixar, lends itself very well for modelling 3D lines. Here first the surface is diced into very small polygons: micropolygons. These are flat-shaded quadrilaterals that are approximately half a pixel on a side. Next these micropolygons are shaded, and finally the polygons are transformed and scan-converted. A z-buffer is used for hidden surface elimination. The principle that calculations should be done in natural coordinates leads to a simple implementation of the model described here: all short-cuts described for precalculation can be applied.

If a scan-line algorithm is used, linear interpolation over the spans can be used for the calculation of the components used in the model. This, however, requires a lot of additional bookkeeping.

11.4 Results

Plate 13 shows the test surface, overlaid with semi-transparent grey grid lines and green contours. Plate 14 shows the use of pseudo-colours to indicate the height, combined with semi-transparent grey grid lines. These images show that the model for 3D lines serves its purpose: the shape of the surface is clearly visible, while simultaneously the position and value of extrema can be located. For analysis purposes simpler and faster techniques could be used, but for presentation purposes this technique serves a need.

Further, the expressiveness of this technique allows to visualize multiple functions simultaneously. Plate 15 shows the test surface again, but here pseudo-colours are used to show a second function, uncorrelated to the height, combined with the same lines as shown in plate 13. The colours were deliberately desaturated, because otherwise they tend to dominate the shading. Plate 16 is a close-up view on the upper left corner of plate 15, which shows the properties of the model. The lines have a constant width in object space, and further they are smooth, curved, shaded and semi-transparent. It is up to the reader to decide whether the illusion of semi-transparent tape stuck on the surface has been realized.

Plate 17 to 19 show some practical applications. Plate 17 shows the wave pattern of a ship calculated by the DAWSON package[8]. The wave heights were scaled with factor 5 for visualization purposes. Parts of the wave surface above zero level are coloured brown, parts below zero level are coloured blue. Further grid lines as well as contour lines are used.

Plate 18 shows a droplet distribution measured by a forward scattering laser spectrometer during the EUROTRAC fog campaign in the Po-valley in 1989. The number of droplets as a function of time (24 hours, starting at noon) and diameter (1mm to 95mm, logarithmic scale) is shown. When fog is formed in the evening, the number of droplets increases rapidly.

Plate 19 shows the distribution of the ^{137}Cs nuclide over the cross section of a Light Water Reactor (LWR) fuel rod (diameter 10.5 mm). The migration of this nuclide to colder areas is clearly visible. For the acquisition of the original data, the fuel rod was positioned in front of a collimator and its radiation was measured with a gamma spectrometer. The distribution over the cross section was calculated from these data with computer tomography.

The images shown here were made with a rendering system developed by the author at the Netherlands Energy Research Foundation ECN. This system is based on the principles described in[4]. The basic geometric element is the bicubic patch, which is diced into micropolygons during the rendering. The lines were calculated procedurally. For the colours in plates 15 and 16 texture mapping was used.

The system was written in C, and runs on a Sun 3/60 workstation. The images were computed at a resolution of 512 *times* 256 pixels. For each pixel four samples were taken and averaged for anti-aliasing purposes. For the final images the surface was diced into about 500K polygons. The execution time of the non-optimized system was about 30 to 45 minutes per image. During the design phase, however, about 1 to 10K polygons sufficed, so acceptable previews could be achieved within minutes.

Colour calculations were done in 24 bits precision (eight per channel). The final step for display was colour quantization in order to reduce the number of different colours to 256. To this end the implementation of J. Poskanzer of the median cut algorithm of Heckbert[7] was used.

11.5 Conclusions

The results show that the visualization of scalar functions over surfaces can be enhanced by modelling lines as 3D objects. The shape of the surface can be judged, while simultaneously the position and value of extrema can be located easily. The additional cues of the lines are particularly helpful if multidimensional data sets have to be shown.

The price for these images in terms of processing time is high, but not prohibitive. Obviously, drawing lines in image space with a fast algorithm such as Bresenham's[6] is much faster, but on the other hand the image quality of 3D lines is superior. Besides an aesthetical improvement, the interpretation of such lines is also more efficient. A underlying physical model, i.e., semi-transparent tape stuck on the surface, helps in the comprehension of the image. The observer can spent his time on the analysis of the data shown, without diversion by their visual representation.

This topic, the visualization of functions over surfaces, has not been exhausted by far. The technique for pseudo-colours is simplistic, the integration of 3D lines with the techniques described in[9] will give better results for multidimensional data. Other real-world analogues, such as grooves or ridges, could be used as a model for lines. In the images shown, the height was used to indicate the value of the function. It could be attempted to use an offset from the surface as a cue for function values for general surfaces, i.e. to show the pressure distribution over an aeroplane as a mountain landscape, mapped on the surface of the aeroplane. Many other techniques are conceivable.

Finally, the viewer of the images is the ultimate judge of the quality of the techniques used for visualization. Therefore, for a better assessment of the quality of those techniques, as well as favourable values for the parameters involved, experiments with test subjects have to be carried out.

Acknowledgements:

I would like to thank J.M. Akkermans, A.R. Burgers, and W.H. Rijnsburger for their critical comments on earlier versions of this paper. I further thank H.C. Raven (Maritime Research Institute Netherlands), B.G. Arends and G. Dassel (both Netherlands Energy Research Foundation ECN) for the pleasant cooperation and their permission to use their data sets.

11.6 References

[1] J F Blinn. Simulation of Wrinkled Surfaces. *Computer Graphics*, 12(3):286–292, 1987.

[2] E Catmull. A Subdivision Algorithm for Computer Display of Curved Surfaces. Technical Report UTEC-CSc-74-133, University of Utah, Computer Science Department, December 1974.

[3] R L Cook. Shade Trees. *Computer Graphics*, 18(3):223–231, 1984.

[4] R L Cook, L Carpenter, and E Catmull. The Reyes Image Rendering Architecture. *Computer Graphics*, 21(3):95–102, 1987.

[5] F C Crow. Summed-Area Tables for Texture Mapping. *Computer Graphics*, 18(3):207–212, 1984.

[6] J D Foley and A van Dam. *Fundamentals of Interactive Computer Graphics*. Addison-Wesley, 1982.

[7] P Heckbert. Color Image Quantization for Frame Buffer Display. *Computer Graphics*, 16(3):297–307, 1982.

[8] H.C. Raven. Variations on a Theme by Dawson. In *Proceedings of the 17th Symposium on Naval Hydrodynamics, The Hague*, pages 151–172, 1988.

[9] P K Robertson and J F O'Callaghan. The Application of Scene Synthesis Techniques to the Display of Multidimensional Image Data. *ACM Transactions on Graphics*, 4(4):247–275, 1985.

[10] E Tufte. *The Visual Display of Quantitative Information*. Cheshire, 1985.

[11] T Whitted. An Improved Illumination Model for Shaded Display. *Communications of the ACM*, 23(6):343–349, 1980.

[12] L Williams. Pyramidal Parametrics. *Computer Graphics*, 17(3):1–11, 1983.

12 Interactive Three-Dimensional Display of Simulated Sedimentary Basins

Christoph Ramshorn, Rick Ottolini, Herbert Klein

ABSTRACT
Simulated sedimentary basins that are represented by volumetric data obtained from a numerical model are visualized by

- a set of isochron sediment surfaces,
- a lattice of vertical sections,
- basement topography, and
- water body.

Sediment surfaces and sections are color-coded to represent sediment composition in alternative ways, including sediment type and sediment age. It is possible to interactively

- show or hide any portion of the basin,
- rotate, resize, and vertically exaggerate the basin,
- highlight a sediment type,
- show basin evolution through time, and
- switch sediment classification methods.

Two viewing programs implemented on top of two different graphics systems (Doré[2] and GL[1]) are described and compared. One viewing application is built from a set of independent but cooperating programs that provide graphic functions and a user interface under X-Windows.

12.1 Introduction

Geologists investigate sedimentary basins because these are prospective oil reservoirs. Sedimentary basins may form where rivers that carry pebbles, sand, and clay (clastic sediments) flow into an ocean. Numerical models can simulate the physical processes that govern transportation and deposition of clastic sediments. SEDSIM (SEDimentary process SIMulation program[7]), for example, generates volumetric data sets representing sedimentary basins. These need to be graphically displayed to evaluate results of simulation experiments.

The following sections first briefly introduce SEDSIM and describe SEDSIM output data. Then we show how we visualize sedimentary basins. Further, we give implementation details of two viewing programs on top of Ardent Computer's Dynamic Object Rendering Environment[2] in conjunction with the Doré User Interface (DUI), and on top of Silicon Graphics' Graphics Library[1]. A short evaluation follows as to the suitability of Doré and GL for our needs. Finally we show how several independent programs comprising graphical controls and a rendering module cooperate to form a viewing application.

12.2 Simulation of sedimentary basins - SEDSIM

SEDSIM represents sedimentary processes by breaking them down into their fundamental components. For example, transportation and deposition of grains of clastic sediment, such as sand, silt, and clay, are represented by the behavior of individual grains swept along by currents of water. Flow of water in streams, or in turbidity currents, or through action of waves in the surf zone, is represented by the mechanics of fluid motion and particle transport. Further processes involve compaction, isostatic compensation, tectonic motion, sea level changes, and so forth.

SEDSIM adheres to the conservation of mass, energy, and momentum. While fluid elements loaded with sediment are tracked in continuous space, deposited and eroded material is accounted for on a grid of rectangular cells of fixed size in X and Y, and of variable size in Z. SEDSIM output essentially describes basement topography and contains a five-dimensional array representing the amount of pebble, sand, silt, and clay deposited in each cell, recorded at user-defined intervals. Typical data sets comprise a grid of 50 by 50 cells with 25 "snapshots" of deposit stages. For efficiency, only cells that have changed between snapshots are stored in the output file.

While experiments currently performed cover areas in the range of several tens of square kilometers and compute some 100,000 simulated years of deposition, accumulated sediment layers pile up to the order of several tens of meters. Usually sedimentary basins show complex sedimentation and erosion patterns.

12.2.1 Display of sedimentary basins

The volumetric information contained in SEDSIM output needs to be visualized three-dimensionally. We regard information clarity as the most important constraint in viewing programs, followed by information amount, and frame generation speed. Speed becomes crucial when basin development through time or fluid flow is visualized in animated sequences.

For rendering volumetric data that are not voxel data there are basically two options. One is to transform the data to voxels and to visualize them by displaying voxel slices, or through perspective display of thresholded voxel blocks, or by rendering isosurfaces. The other option is to find geometric shapes which provide a suitable graphic representation of the data. SEDSIM output has two properties that make it little suited for voxel representation: (1) Vertical thickness of deposits is orders of magnitudes less than their lateral extension (plate 22); (2) vertical resolution is much finer than lateral resolution. Besides it is preferable to provide graphic elements that are familiar to end users, in this case to geologists.

We represent three-dimensional sedimentary basins as simulated by SEDSIM as

- a set of isochron sediment surfaces (plate 20),
- a lattice of vertical sections (fence diagram, plate 21), and
- basement topography (plate 22).

The sediment surface and the fence diagram are color-coded to represent sediment composition in a number of alternative ways, such as classified by sediment type (pebble, sand, silt, clay; plate 25), or by sediment age (plate 26). Colored surfaces resemble geologic maps that are projected on a digital elevation model; fence diagrams are traditional graphical representations of subsurface structures. Deposit surface and interior can be viewed simultaneously if the surface is rendered translucent (plate 24). Basement topography can be optionally displayed to visualize basement erosion (plate 22).

In sediment type classification, colors are specified in the RGB (red, green, blue) color model[3]. Red represents coarse sediment, blue fine sediment and green in between. Sediment compositions are mapped to mixtures of red and green or to mixtures of green and blue. Ternary mixtures are not used because they tend to produce grayish hues that do not present much information. In sediment-age classification, the HSV (hue, saturation, value) color model[3] is used. Sediment of each time step is assigned another color by sampling the HSV color circle at user-defined intervals. While smooth transitions between subsequent colors are helpful for visualizing discontinuities (plate 26), more distinct colors for each time step visualize layer boundaries better (plate 27).

A translucent blue box outlines the water body above the sediment (plate 23). When displaying both the fence diagram and water, the water can be prevented from "flooding" the space between fences that reach above the water surface by rendering the sediment surface completely transparent before drawing the water. Thus, only the z-buffer and not the display will be altered when drawing the sediment surface. Then water is only displayed where the sediment surface drops below the water surface.

Basin evolution through time is visualized by animating a sequence of snapshots. This is accomplished by automatically rendering one snapshot after the other. Clearly, animation depends on frame generation speed, which is currently between 0.5 and 3 seconds depending on the amount of data rendered. Therefore, and because a basin can considerably change between snapshots, animated sequences may appear jerky.

Interactive control of the image manages the information content by hiding or highlighting basin components, or by changing viewing angle. Graphical controls appear as button sets or popup menus, as dial sets, and as vector definition controls (plate 24). Buttons and popup menus set discrete states such as hiding or displaying basin elements. Dials adjust continuous states, for example viewing angle or vertical exaggeration. Vector definition controls let the user define light source position and switch lights on or off.

Much of the interactive display control consists of changing the visibility of various basin elements. This both controls information complexity and increases speed when the number of display elements is reduced. For example, it is feasible not to display too many sedimentary fences at a time to prevent images from getting cluttered.

Vertical exaggeration helps to view sediment layers that are otherwise too thin to present any information (plate 22). There is an option to exaggerate sediment layers without exaggerating the other components of the basin. This prevents the basin display from being rendered as a high but thin column.

Limited interactive reclassification is accomplished by changing the illumination model. Changing the light source(s) from default white to non-white highlights a particular image hue and sediment type (plate 27).

Two viewing programs are implemented, SEDSHO and SedView. SEDSHO runs on an Ardent TITAN computer using Ardent's Doré User Interface (DUI), which in turn is built on top of Doré and X-Windows[5]. SedView runs on Silicon Graphics 4D series workstations with Silicon Graphics' graphics library GL, and an user interface also built upon X-Windows. Both computers run under Unix V.3. Plates in this paper have been produced with SedView.

12.3 SEDSHO (using Doré and the DUI)

Doré is a full-featured three-dimensional graphics system supporting user interaction. It sees graphic primitives and graphic operations (such as scaling, clipping, shading, etc.) as objects in a hierarchic database. It has a large set of graphics primitive objects including points, lines, polygons, triangle meshes (for surfaces), basic solids, and text. Color, size, viewing angle and lighting model attributes are specifiable. Images are created by grouping these graphic objects along with graphic operations into a database. Graphic objects are rendered by applying perspective transformations and shading while traversing the database. Several functions allow modification of the traversal path providing an efficient method for interactive scene manipulation. Visibility of selected objects, for instance, can be manipulated by functions that switch pointers in the database.

The User Interface program (DUI) consists basically of a prebuilt Doré database and an interactive control panel implemented on top of X-Windows. The database contains function objects setting up a "studio", a "camera", "lights", and so on. These objects can be manipulated through the control panel. The panel entails buttons and dials that are configurable to produce specific commands. A programmer can extend the DUI by adding his own data objects and more graphic function objects to the prebuilt database. He further may supply code and user defined commands to extend the interactive functionality of the DUI. SEDSHO has been implemented this way.

The main graphics primitive used in basin display with Doré are triangle meshes with vertices that are defined by location and color. Basement topography, fences, and deposit surface are transformed to groups of triangle meshes. One group is built for each snapshot. Newer versions of Doré provide functions for rapidly updating locations and colors of triangle mesh vertices while maintaining mesh topology (the triangulation pattern). Basement and surface mesh objects could be built once and then updated for each snapshot. However, fence meshes would need to be entirely recomputed either way to account for triangulation changes caused by erosion. The fact that all meshes are currently computed at program start time increases interactive speed and decreases program complexity. When the number of rendered triangles increases with data set size, memory limits may become a serious constraint. Less commonly viewed surfaces will then have to be computed when they are requested.

In the current version of SEDSHO, all of the sediment classifications are also computed at program start time because there is insufficient memory to keep both SEDSIM data and the Doré database at the same time.

12.4 Sedview (using GL)

GL, as opposed to Doré, is more a low level graphics library that includes procedures for 3D rendering. Programmers must maintain their own databases for the scenes and call appropriate graphics functions with user supplied data.

SedView builds a compact database from SEDSIM output. This data base stores initial basement topography and incremental deposit changes (differences to previous stages rather than absolute values are stored) for each snapshot. Incremental cell changes are coded as one-byte values that express cell thickness as a fraction of the maximum thickness difference produced during the previous snapshot interval. Grain size distribution within a cell is expressed by three bytes for storing the percentage of each grain size. While this approach minimizes storage requirements, each deposit stage must be constructed from the database prior to rendering. SedView provides extra storage for one complete snapshot description including basement topography, elevation and color of each cell, and water table elevation. GL's rendering functions are called with these data.

Basement topography and deposit surface are rendered as shaded rectangle meshes. Fences are drawn as triangle meshes as in SEDSHO. GL imposes a limit of 256 vertices per triangle mesh but with current data set sizes, this presents no obstacle.

GL requires that for each vertex of both surface meshes and fence meshes a normal vector is supplied. Surface mesh normals are computed by averaging the normals of the four triangles that can be built from a vertex and its four direct neighbours (or one triangle at each corner vertex, or two triangles at each vertex along the sides of the mesh, respectively). Fence vertex normals are defined to point upwards, with a slight bias to -X in fences that parallel the Y-axis. Thus fences parallel to the X-axis and fences parallel to the Y-axis can be rendered with slightly different brightness even when using only one light source (plate 25).

SedView lets the user specify light source position. Lights that are positioned at low angles with respect to deposit surface visually enhance topographic features.

12.4.1 Doré and GL

Doré, especially in conjunction with the DUI enables programmers to visualize their 3D data sets efficiently. The strength of Doré is best illustrated by the fact that even novice users can quickly accomplish visualizing their data three-dimensionally. This is due to the high level approach of Doré, and to many helpful details in its design. For instance, it provides functions that automatically transform user data into studio space guaranteeing that something will be displayed. Novice users of other 3D graphics systems often need multiple try and error cycles before they succeed in making their display objects appear on the screen.

The cost of the this approach is considerable memory consumption and overhead for building the database. When it comes to operations that are not supported by Doré, for instance more complex object editing, two databases must be maintained simultaneously. This may lead to extra memory, computing time, and programming effort.

Systems like GL require more coding right from the beginning, e.g. for scenery setup, or for maintaining a display list, or for supporting user interaction. On the other hand, these systems can be better tailored to specific needs of an application.

Animating series of SEDSIM snapshots may serve as an example. In the Doré version, for each snapshot a group of objects describing the complete scenery must be stored. Animation is performed by rendering group by group. Since database object generation for each set of fences requires some 10 seconds, for animation each time step must be prebuilt. With the current practice of precomputing the meshes for each snapshot and each classification, the number of displayed time steps is limited to about 25, given a 50 by 50 grid on a computer with some 130 megabyte virtual memory. In the GL version, incrementally computing one snapshot requires less than a second for a 50 by 50 grid on a 10 mips machine. Roughly the same frame generation speed is achieved at considerably less memory use (about 2.5 Mbytes). For similar reasons switching data sets requires up to a couple of minutes in the Doré version and about 5 seconds in the program using GL.

12.5 User interface

SedView is implemented as a set of independent programs that cooperate via UNIX interprocess communication. SedView modules include (figure 12.1)

- sedview, a command driven viewing program;
- geo3filer, a file handling and I/O redirection utility;
- dials ("Transform"), a graphic control generating commands with scalar parameters;

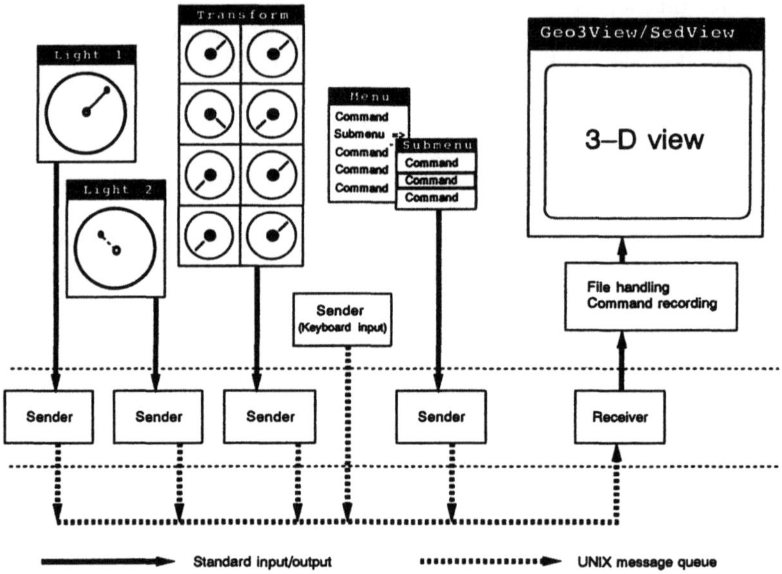

FIGURE 12.1. Scheme of cooperating processes. Processes visible to the user (upper level) are connected (middle level) to an interprocess communication mechanism (lower level)

- ibutton ("Menu"), a control providing popup menus;
- sphere ("Light 1"), a graphic control generating commands with vector parameters;
- geo3send, connecting its standard input to a specified message queue;
- geo3receive, connecting a specified message queue to its standard output.

Sedview is the actual viewing program; Geo3Filer intercepts special commands and provides file handling support including a command recording/replaying mechanism. Recordings of user actions can be generated, edited, and be used as scripts for animated sequences. All user interface components are configurable by resource files to produce specific sets of commands.

The user interface is implemented on top of X-Windows using some ideas realized in other software packages. Plate 24 shows a dial control (labeled "View parameters") with eight "knobs" for giving values such as rotation angle and viewing distance. The basic layout for this control we saw first in the DUI.

Another type of control shown in plate 24 (labeled "Light 1") serves to define vector direction and is used for setting light source position and switching lights. To understand how it works, imagine a pin (small black circle = pinhead) sticking in a small sphere (small circle in the center). Moving the pinhead refers to pointing to different directions. The "pinhead's" position is transformed to vector components. This control is built after a component of the ConMan environment[4].

The third type of control (labeled "Sedview Commands", plate 24) generates commands according to user choices from popup menus. Currently this control is not implemented under X-Windows but uses popup menu functions provided by GL.

Figure 12.1 illustrates how components of three levels work together. The user deals with graphic controls and watches 3-D rendering output (upper level). Background sender and receiver processes connect standard input/output channels of involved components via UNIX pipes (middle level). Senders and receivers communicate with each other by UNIX message queues (lower level). A sender process invoked from the keyboard rather than as part of a pipe (sender process in upper level) serves to connect the keyboard to the receiving process.

The approach outlined above has several advantages. First, each module can be developed and tested separately. In fact, the viewing module and the user interface were developed independently and merged when they were finished. Second, each module can be kept as simple as possible. For example, the rendering component is command driven and file handling is performed external to the rendering program in a separate module. Third, modules can easily be added. As soon as a new module becomes available, e.g. a new graphic control, it can be used in the system. Fourth, modules can be reused in other configurations. In fact, SedView is an extension to Geo3View which is a viewing program for geologic structures represented by triangle meshes[6]. Fifth, modules only need to use their standard input/output streams to communicate with other modules. Standard piping facilities available under UNIX can be used to configure the viewing application, including connections across a network via remote shells.

The advantages listed above must be paid for with interprocess communication overhead. There is a tradeoff between gain of flexibility, and both lesser data transfer efficiency and additional effort for programming and maintaining communication modules. A fine example of a sophisticated graphic environment for an "orchestrated collection" of modules is given by Haeberli[4]. Our implementation of cooperating processes is quite limited, but it provides considerable flexibility. This is important in the university environment where many students may contribute modules that need to be integrated into larger applications.

12.6 Future directions

Planned improvements of the viewing programs include visualizing flow as animated particles, and visualizing migration paths (connected regions that have a given minimum permeability). The generated images need to be refined; for example, when coloring the surface mesh, colors are smoothly interpolated between adjacent vertices. This leads to somewhat smeared surface colors that are not appreciated by geologists who are used to maps with discrete borders between mapped units. A solution to that might be to classify sediments at the deposit surface and insert pairs of additional vertices to model boundaries between discriminated classes.

As SEDSIM is evolving, more modules will be added. They need to be tested separately as well as in interaction with other modules. There is need to visualize how each process affects an evolving sedimentary basin.

It is desirable that the task of setting up an experiment configuration is graphically supported, for instance by pointing and clicking to specify data paths among computation modules. Users should be able to graphically monitor data that are exchanged between modules.

Acknowledgements:

Visualization programs have been developed in cooperation between the Geomathematic Program of the Department of Applied Earth Sciences (SEDSIM project) at Stanford University, California, and the Geo3D project of the Geological Institute at Freiburg University, West Germany. We thank John W. Harbaugh (Stanford) and Reinhard Pflug

(Freiburg) for making this work possible, and for their advice.

The SEDSIM project is sponsored by several oil companies. The Geo3D project was funded by the German Research Foundation. Silicon Graphics loaned a Personal Iris workstation.

We also wish to thank Young Hoon Lee, Paul Martinez, and Johannes Wendebourg for providing data examples and user feedback. We are grateful to staff at both Ardent Computer and Silicon Graphics for their support.

12.7 References

[1] *GT Graphics Library User's Guide.* Mountain View, CA, USA, 1988.

[2] *Dore Programmer's Guide.* Sunnyvale, CA, USA, 1989.

[3] J D Foley and A van Dam. *Fundamentals of Interactive Computer Graphics.* Addison Wesley, 1983.

[4] P Haeberli. ConMan: A Visual Programming Language for Interactive Graphics. *Computer Graphics*, 22:103–111, 1988.

[5] A Nye. *X-lib Programming Manual for Version 11.* O'Reilly, Newton, Mass, USA, 1988.

[6] Ch Ramshorn, H Klein, and R Pflug. Dynamic Display for Better Understanding Shaded Views of Geologic Structures. *Geol. Jb. Hannover (in press)*, 1990.

[7] D Tetzlaff and J W Harbaugh. *Simulating Clastic Sedimentation.* Van Nostrand Reinhold, New York, USA, 1989.

13 Visualization of 3D Scalar Fields Using Ray Casting

Andrea J.S. Hin, Edwin Boender, Frits H. Post

ABSTRACT
In this paper a high-quality rendering technique for the visualization of 3D scalar fields in an "aquarium model" is presented. This technique is based on ray casting. The model consists of scalar values defined on a grid, supplemented with glass walls and a bottom. Using trilinear interpolation within the grid is an option. The implementation has been divided into two stages; ray casting, and colour binding using transfer functions are implemented as separate processes. This enables the user to generate different pictures from the same viewpoint, experimenting with different parameters. The method has been applied to data obtained from simulations computing concentrations in sea water.

13.1 Introduction

In recent years, interest has grown in computer-generated visualization of 3D data obtained from measurements and numerical computations[4]. The enormous increase in power and availability of computing resources has stimulated the development of numerical models of ever growing scale and complexity. Manual interpretation of numerical data, always a tedious task, has become virtually impossible. For the further development of measurement and analysis techniques, there is an urgent need to exploit the high bandwidth of human visual perception in grasping the spatial structure of complex phenomena. Therefore, new visual data presentation techniques are being developed. Especially in the case of 3D simulations, traditional 2D visualization often fails to provide the necessary insight, especially when an overall view of spatial structure is desired.

In visualization of 3D data, the aim is to promote a primarily qualitative understanding of the global spatial structure. 3D data from numerical simulations or measurements are often represented as scalar or vector fields, defined by scalar or vector values on a regular 3D grid. Visualization of these fields can be supported by showing 3D objects or shapes representing the context, and 2D sectioning or slicing is provided for closer study in selected planes. With 2D visualization techniques, data are then shown in one or more projections or cross sections, using contour lines, flow lines, and similar aids to allow a more quantitative view of the data.

This paper describes a visualization technique for 3D univariate scalar fields, intended to show the spatial distribution of a diffuse field in a transparent medium. The applicability of the method is demonstrated in the study of the distribution of concentrations of silt or chemicals in sea water. Data for this were supplied by Rijkswaterstaat, the Dutch State Agency for Public Works. The data were computed concentrations, obtained from 3D numerical flow simulations, and represented as 3D scalar values on a regular rectangular grid, supplemented by depth measurement data.

The method uses ray casting to generate images of a so-called aquarium model (see figure 13.1). The region to be studied is represented as a rectangular tank, bounded by transparent glass walls, and an opaque surface for the sea bottom or the coast. The distribution of the scalar field of concentration values is shown as coloured diffuse clouds inside the aquarium. The data of the scalar field over the 3D volume is conveniently represented as a 3D array of samples. The spatial extent of the volume elements, located

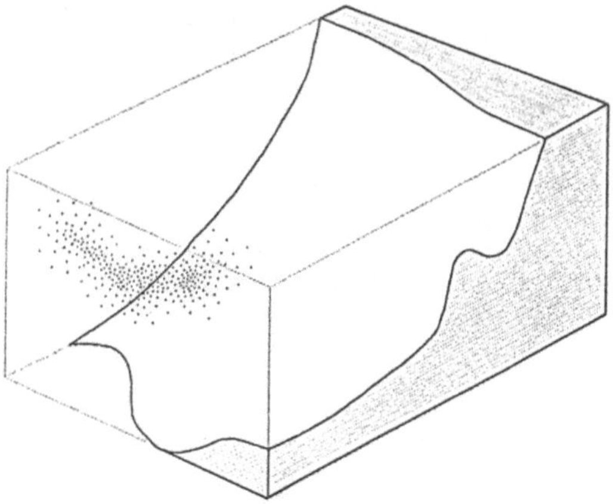

FIGURE 13.1. The aquarium model

inside the aquarium model, is referred to as the data model.

The method is related to other volume visualization techniques based on ray casting, such as those described by Upson and Keeler[7] and Sabella[5]. Because of the transparent medium, in which the scalar field is visualized by a gradual decrease of transparency, a front-to-back volume rendering technique is generally preferable to the simpler back-to-front rendering techniques[3].

Upson and Keeler[7] describe two rendering techniques that use volumes as the basic geometric primitives. These techniques use a linear approximation of the scalar field within each volume element, called a computational cell. The first technique is based on ray casting, and processes the cells encountered by a ray emanating from the viewpoint. The second technique, called cell-by-cell processing, is a cell- oriented front-to-back method. A cell is projected onto the pixels of the screen starting with those on the plane closest to the viewpoint. The contribution of each cell is accumulated for each pixel.

Sabella[5] also uses ray casting to generate images showing certain properties of a scalar field, e.g., the distance to the peak value along a ray, or its center of gravity. These properties are mapped to HSV colour space to produce an image.

The method described here is based on the first method described by Upson and Keeler, adapted for application with the aquarium model; we have extended it using flexible methods for assigning colours to the field values, and we have implemented it as a two-stage process, allowing the user to experiment with several colour variations.

We first describe our ray casting technique, and then the final assignment of colours to the scalar field values. Next, the implementation as a two-stage process is described, and we give some results and pictorial examples. We end the paper with a brief discussion and directions for further development.

13.2 Ray casting

Ray tracing is a widely used technique in computer graphics to generate high-quality images incorporating optical effects such as reflection, transparency and shadows. In conventional ray tracing, rays are sent from the viewpoint through each pixel of the screen into model space, and intersections are calculated with surfaces hit by the rays. From the

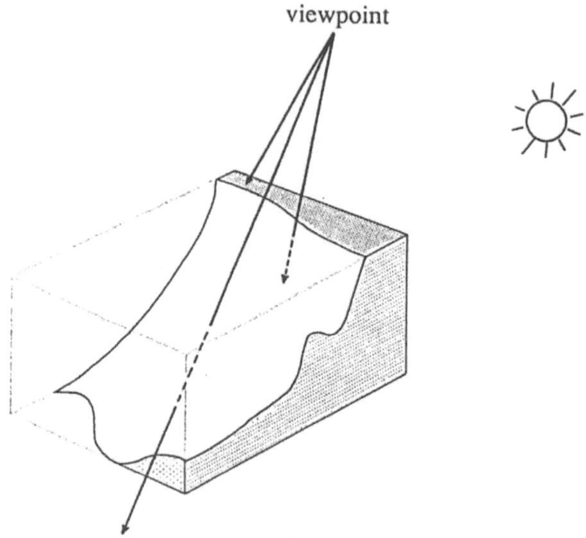

FIGURE 13.2. Ray casting

first intersection point, secondary rays can be traced in the direction of reflection from the surface or refraction into transparent material, or go in the direction of a light source. These secondary rays serve to simulate the special optical effects.

In the present case, we use a simple type of ray tracing called ray casting. No secondary rays are cast, and no mirroring, refraction or shadowing are determined (see figure 13.2). Rays are not only intersected with the surfaces of the aquarium model, but also cast into the data model, where scalar values are integrated along the rays, just as a light ray is attenuated when it penetrates a semi-opaque liquid. A ray ends when an opaque surface (the sea bottom in the aquarium model) is encountered; there, an intersection point is calculated and light reflection is computed as in conventional ray tracing. A ray also ends when it leaves the model.

To determine which cells are intersected by a ray, the simple and fast voxel traversal algorithm of Amanatides and Woo[1] is used. Since the domain is partitioned into cells of uniform size, traversing the model is easiest in data model space, where a simple relation exists between a point in the data model and the cells containing the scalar values needed to compute the scalar value at this particular point. Determination of this value depends on the type of interpolation desired: for constant-valued cells, the value in the centre of the present cell is sufficient, but for trilinear interpolation of the scalar field within a cell, the values of the eight nearest cells are needed (see figure 13.3).
For constant-valued cells (voxels), the integrated value I of the scalar field along a ray is determined by:

$$I = \sum_i c_i \cdot x_i$$

with c_i the scalar value in the centre of the i-th voxel along the ray and x_i the distance along the ray inside the voxel.

For trilinear interpolation, the basic formula remains the same, but now c_i is defined as the mean of the scalar values at the cell transitions ($c_{i,in}$ and $c_{i,out}$), and x_i as the distance

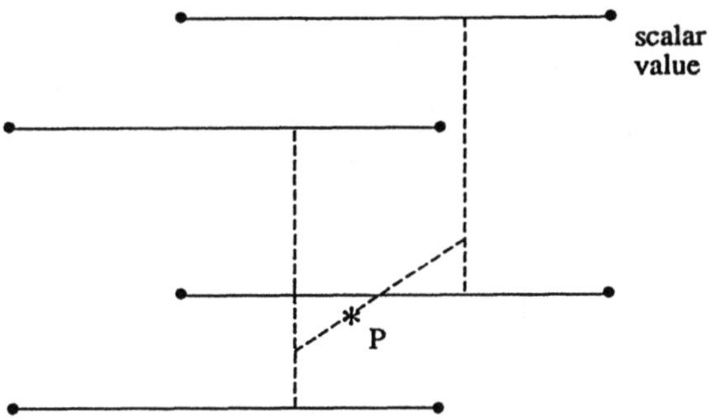

FIGURE 13.3. Trilinear interpolation for P using eight scalar values

along the ray through the cell:

$$I = \sum_i \frac{1}{2}(c_{i,in} + c_{i,out}) \cdot x_i$$

13.3 Colour mapping and image generation

A colour of an element is specified using the HSV colour model, which is more suitable for user input than e.g., direct specification in RGB components[6]. Different scalars can be mapped to opacity or to different components of colour space, using a transfer function which performs this mapping. The functions, or choice of the parameters of these functions, can be varied to obtain different effects.

As implemented now, the integrated value of the scalar field along the ray is mapped to the opacity of the field. The opacity of an element must be specified as a value in the range 0 to 1, and indicates the contribution of an element to the final colour of a pixel. The transfer function used is linear, though an exponential function (see figure 13.4, continuous line) would probably be more appropriate and in conformity with physical reality. For opacity a default scaling is used that maps the global maximum of the integrated values along the rays to an opacity of one. A compression (or scaling) factor can be used to show a more or less attenuated picture. The user can supply the compression factor, and the resulting values are simply scaled by this factor (see figure 13.4).

The opacity indicates the contribution of the colour of an element to the final colour of the pixel. The colour C of a pixel is computed as a weighted sum of colours and opacities of the elements along the ray:

$$C = \sum_E CE \cdot OE$$

with E the elements in order of appearance along the ray (first glass wall, scalar field, bottom, second glass wall); CE is the colour of an element, and OE is the opacity of each element. The summation stops when $S \cdot OE = 1$. If, after traversing the whole model, $S \cdot OE < 1$, the remaining fraction is filled with background colour. Since colours cannot be merged correctly in HSV-space, they are transformed to RGB- space before being merged. Furthermore, the opacity of a glass wall can be adjusted by changing the contribution of its colour to a pixel. As seen above, this will also have an effect on the contributions of the other elements.

It is also possible to map the integrated value to hue, to obtain a gradual variation in colour. The user must select a range of colours, and the integrated value will be mapped

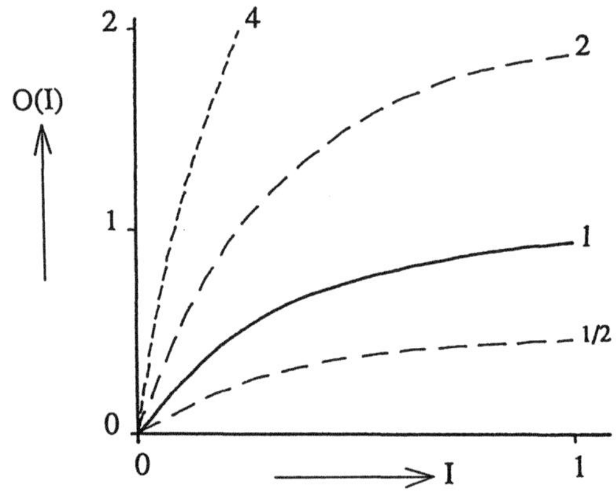

FIGURE 13.4. Exponential transfer function between the integrated value I of the scalar field and the opacity O(I) for different values of the compression factor

to the hue of these colours. Depth cueing is realized by a reduction of the V-component of the element where traversal has stopped. For this the length of each ray is required.

13.4 Implementation

Our implementation is divided into two stages. The first stage, during which rays are cast through the model, is viewpoint dependent, is computationally most expensive, and serves as a preprocessing stage. The information generated in the first stage for each pixel is used in the second stage, where a mapping is established between the colour-independent information and user-supplied, colour-dependent parameters. At that stage a colour is assigned to each pixel. As this second stage is much faster than the first, the user can experiment with different parameters to establish a mapping between volume data and colour (using the same viewpoint).

During traversal of the cells, information about which of the elements were encountered by a ray (walls, bottom or scalar field) are stored in a file using an encoded number to indicate which elements were found. Next, information relevant for visualizing these encountered elements in the second stage is stored. As for faces (e.g., walls and bottom) the diffuse reflection of light on a face is needed, the cosine of the angle of incidence of the light is calculated and stored for each element. To visualize the scalar field, the integrated value of the field along the ray is calculated and stored. Additional information about the length of the ray is stored to generate images with depth cueing.

After the information for all rays (pixels) has been calculated, general information is gathered and stored. During ray casting of the entire data model, the maximum integrated value of the scalar field of all rays is determined, to enable the adjustment of the opacity of the field in the second stage. Another part of the general information is the range over which all ray lengths vary to correctly adjust depth cueing.

Since the information stored for each ray is independent of any material properties (such as colour or transparency) of the elements encountered during ray casting, different images can be generated from the intermediate file for various visualization parameters and mappings.

13.5 Results

Programs for both stages have been written in C and run on a VAX-11/750 with floating point accelerator, using the Unix[1] 4.3 bsd operating system. Images were displayed on a Pluto Colour Graphics Display at a resolution of 768 x 576, with 24 bits of colour per pixel.

Plates 29 to 31 show a region of study just below sea-level. It represents a part of the Dutch coast between the latitudes of Amsterdam and Rotterdam. The dimensions of the region are 15 kilometers wide, 55 kilometers long, and only 20 meters deep. The depth of the model has been scaled up to allow full use of the three-dimensionality of the visualization technique. The data model consists of 15 x 55 x 5 cells. The scalar field represents the computed distribution of silt some time after a simulated dump by a mud-barge. In all images a linear transfer function to opacity is used.

In plates 32 to 34 a larger region of study of the same coastal region is shown with a schematic channel in the direction of Rotterdam. The dimensions are 92.8 kilometers wide, 134.4 kilometers long, and 24 meters deep. The data model consists of 58 x 84 x 4 cells, with each cell corresponding to a volume of 1600 x 1600 x 6 meter in reality. The cloud represents a computed scalar field of concentrations caused by a polluting source put at the bottom in the model. A linear transfer function to opacity is used in all images,with linear mapping to a colour range in plates 33 and 34.

Ray casting proved to be quite expensive; total processing times for the pictures of plates 29 to 31 were between 1 1/2 and 2 hours, for the pictures of plates 32 to 34 between 2 and 3 1/2 hours. This is mainly caused by the traversal of the cells, including the integration of field values. The preprocessing time directly depends on the number of cells traversed per ray. Also, a substantial part of the time is used for displaying the geometry of the environment, i.e. intersecting the rays with the walls and bottom.

The difference in image quality between constant-valued cells and trilinear interpolation was smaller than expected (compare plates 33 and 34). It is doubtful whether the interpolation is worth the extra computational cost of about 30%.

The division into two stages proved reasonably successful. Preprocessing takes up most of the time (more than 85%), and in the second stage the user can generate an image in 7 to 15 minutes. The main disadvantages are that the viewpoint cannot be changed at this stage, and that model data are not available for other visualization techniques.

Engineers of Rijkswaterstaat, who hitherto had only experience with techniques for visualization of data in 2D planes, were enthusiastic about the resulting images. They welcomed the availability of an overall view of the phenomenon, and also appreciated the power of colour in visualizing data.

13.6 Discussion

Visualization of the aquarium model by ray casting is an example of high-quality volume rendering. Taking many samples for each cell results in smooth variations in colour and opacity, and a convincing effect of a diffuse field suspended in water is produced, suitable for intuitive interpretation of spatial distribution. The effect is supported by showing the glass walls and the bottom of the model. The pictures of plates 29 to 34 closely resemble an "artist's impression" that was manually rendered at the beginning of the project.

The present implementation has not been optimized for speed. Processing times for ray

[1]UNIX is a tradmark of A T & T

casting and image generation can be reduced in several ways. The display of the walls and bottom of the aquarium could be generated separately, using polygon rendering methods, and merged with the scalar field later.

Ray casting could be speeded up by using parallel rays (in effect placing the viewpoint at infinity, resulting in a parallel projected image), which would greatly simplify both cell traversal and depth calculations. Also, it would be worthwhile to use adaptive subsampling in screen space[2] to reduce the number of rays cast.

Several extensions of the method and the implementation are possible. Interactive facilities can be added for selecting projections and cross-sections, so that the 3D method can be combined with traditional 2D visualization techniques. For this, the system can be linked to an existing 2D visualization system. Wire frame previewing would be useful for specifying the viewing parameters for ray casting. Non-linear transfer functions can be made available to the user, and the interpretation of the scalar fields as proposed by Sabella[5] can be added. The method may also be adapted for visualizing multi-variate scalar fields, and for other types of applications, such as atmospheric phenomena.

Finally, there is the question whether results of comparable quality can be achieved using fast back-to-front volume rendering methods. Because each cell is treated as a single data item in these methods, it will be difficult to achieve good transparency effects; also, the aliasing caused by taking only one sample per cell will hamper a good display of diffuse fields. In ray casting, the resolution of the image and the volume data are independent, so that many samples can be taken for each cell, and good visual effects can be achieved. It remains an open question whether fast volume rendering methods can be adapted for this. The ultimate resolution of this question will also depend on the application and the users who must be willing to pay the extra cost for high-quality pictures.

Acknowledgements:

This work was carried out as the first author's engineer's thesis project, with the other authors acting as her advisers. Thanks are due to Wim Bronsvoort, Johan Dijkzeul, Erik Jansen, and Denis McConalogue for their valuable comments on earlier versions of this paper. Special thanks are also due to Johan Dijkzeul from ICIM, who also acted as an adviser, to the people of the Tidal Waters Division of Rijkswaterstaat, for making available the data sets, and to Tjark van den Heuvel of Rijkswaterstaat, whose artist's impression provided the initial inspiration for the development of the aquarium model.

13.7 References

[1] J Amanatides and A Woo. A Fast Voxel Traversal Algorithm for Ray Tracing. In G. Maréchal, editor, *Proceedings of the Eurographics-87*, pages 3–10. North Holland, August 1987.

[2] F Bronsvoort, J V van Wijk, and F W Jansen. Two Methods for Improving the Efficiency of Ray Casting in Solid Modelling. *Computer Aided Design*, 16(1):51–55, 1984.

[3] G Frieder, D Gordon, and R A Reynolds. Back-to-Front Display of Voxel-Based Objects. *IEEE Computer Graphics and Applications*, 5(1):52–60, 1985.

[4] B H McCormick, T A DeFanti, and M D Brown. Visualization in Scientific Computing. *Computer Graphics*, 21(6), 1987.

[5] Sabella P. A Rendering Algorithm for Visualizing 3D Scalar Fields. *Computer Graphics (Proc. Siggraph 88)*, 22(3):51–58, July 1988.

[6] A R Smith. Color Gamut Transform Pairs. *Computer Graphics (Proc. Siggraph 78)*, 12(3):12–19, July 1978.

[7] C Upson and M Keeler. V-Buffer: Visible Volume Rendering. *Computer Graphics (Proc. Siggraph 88)*, 22(3):59–63, July 1988.

14 Volume Rendering and Data Feature Enhancement

Wolfgang Krueger

ABSTRACT
This paper describes a visualization model for 3D scalar data fields based on linear transport theory. The concept of "virtual" particles for the extraction of information from data fields is introduced. The role of different types of interaction of the data field with those particles such as absorption, scattering, source and colour shift are discussed and demonstrated.
Special attention is given to possible tools for the enhancement of interesting data features. Random texturing can provide visual insights as to the magnitude and distribution of deviations of related data fields, e.g., originating from analytic models and measurements, or in the noise content of a given data field. Hidden symmetries of a data set can often be identified visually by allowing it to interact with a preselected beam of "physical" particles with the attendant appearance of characteristic structural effects such as channeling.

14.1 Introduction

Scientific measurements or model simulations typically create a huge amount of field values on a set of discrete space points. Storing this information as large amounts of printed output or tapes often impedes a quick evaluation of the results and an estimate of their scientific value. In order to overcome this obvious bottleneck, tools for the (interactive) visualization of such data fields have been developed over the last few years (see the general discussion of this problem in[12]). Success in the application of such tools has been demonstrated in fields such as astrophysics, meteorology, geophysics, fluid dynamics, and medicine. Generally, in all visualization tools suitable to scientific applications there is a trend to incorporate results and methods from "neighbouring" areas such as pattern recognition, picture processing, computer vision, theory of perception, scattering theory, and remote sensing.

The aim of this paper is to develop special tools for the visualization of 3D scalar data fields originating from scientific measurements or model simulations by supercomputers. The approach will be based on the linear transport theory for the transfer of particles in *inhomogeneous amorphous media*. The advantages of this model are its rigorous mathematical formulation, the applicability to data sets originating from different fields such as molecular dynamics, meteorology, astrophysics and medicine, and a wide variety of possible mappings of data features onto the model parameters. But, visualization based on this model is relatively time-consuming, especially in cases where non-trivial scattering processes are considered. It can be shown that almost all volume rendering techniques, more or less dedicated to the problem of interactivity, are covered by certain mappings and approximations of this model. A discussion of relevant volume visualization models can be found in [9].

The discussion of the volume rendering model proposed in the paper divides into the following main parts:

- Introduction of the concept of "virtual" particles interacting with the data field. By this one is to imagine probing the data set, considered as a 3D abstract object, with a beam of fictitious particles whose properties and laws of interaction with the data set are chosen at the discretion of and for ease of interpretation of the user. Information about the data set is visually extracted from the pattern on the screen of these "scattered" virtual particles. Classical transport theory provides the quantitative framework in which this concept of "virtual" particles can be systematically developed and exploited.

- Development of a mathematical-physical framework to guarantee flexibility in the rendering process for a broad variety of data sets orginating in widely diverse fields. It is desirable to have as many conveniently tunable parameters as possible built directly into the algorithm. Classical linear transport theory with its scattering cross sections, absorption coefficients, internal and external source terms and energy shift term is a familiar formalism whose results are easily interpreted after a minimal amount of "working in" orientation.
- Additional improvements such as texture rendering can be used to probe fluctuations in the data set or to enhance deviations of two related sets. "Interference" patterns visible among scattered artificial particles can be used to identify periodicities and similar hidden symmetries or to indicate the localization of "hot spots" of the probed data field.

The applicability of results and methods of transport theory in the field of computer graphics is well-known: enhanced ray tracing algorithms[15], rendering tools for volumetric effects such as haze or clouds[2, 14, 19] and radiosity methods[11].

In the next section a brief introduction to an appropriate form of the basic equation of transport theory is given. An overview of the numerical computation routines is outlined in the appendix.

In the following section the mapping routines of special data features onto the parameter fields of transport theory are explained. The "physical" action of these visualization parameters is documented with test pictures.

The last section is dedicated to tools which can enhance the perception of data field features. Random texturing is introduced as a tool for comparison of data fields originating from different sources, e.g., analytic solutions and results of experiments, or for visualizing the noise content of a data set. Hidden symmetries in a large (noisy) data field are visualized with the aid of interacting "virtual' particles which can show characteristic "channeling" effects, for example.

14.2 Basic technique for volume rendering: the transport theory model

The visualization model considered follows the concept of extracting the essential content of a 3D data field by "virtual" particles passing the field. The expression "virtual" describes the fact that for visualization applications the particles can interact with the field according to relevant physical laws or artificially chosen ones. The concept of "virtual" particles generalizes the models for tracing light rays in complex environments used for computer graphics applications, where the interaction of the light with the objects is governed by optical laws.

The fundamental quantity in transport theory is the intensity $I(\mathbf{x}, \mathbf{s}; E)$ describing the number of particles at a point $\mathbf{x}$ which move into direction $\mathbf{s}$ with energy ("colour") E. In the discrete colour space the intensity is given by the averaged values $I_i(\mathbf{x}, \mathbf{s})$ with $i = R, G, B$.

The rendering techniques proposed is based on an evaluation of the linear transport equation described in many textbooks (see e.g., [16, 4, 13, 7]). The basic equation of stationary transport theory is the linear Boltzmann equation describing the gains and losses of the particle intensity in a volume element. A form suitable for the visualization of 3D data fields is given by

$$\begin{aligned}(\mathbf{s}\cdot\mathbf{D})\cdot I(\mathbf{x},\mathbf{s};E) &= -\sigma_t(\mathbf{x};E)\cdot I(\mathbf{x},\mathbf{s};E)+q(\mathbf{x},\mathbf{s};E)\\ &\quad -S_{in}(\mathbf{x})\cdot\frac{\partial I(\mathbf{x},\mathbf{s};E)}{\partial E}\\ &\quad +\sigma_s(\mathbf{x};E)\cdot\int d\omega'\cdot p(\mathbf{x},\mathbf{s}'\rightarrow\mathbf{s})\cdot I(\mathbf{x},\mathbf{s}';E)\end{aligned} \tag{14.1}$$

where $\mathbf{D}$ is the gradient operator such that $(\mathbf{s}\cdot\mathbf{D})=\frac{\partial}{\partial R}$ with the distance variable R and $d\omega'$ is the solid angle around the direction $\mathbf{s}$ of the path.

The relevant parameter fields for the visualization process are the extinction coefficient $\sigma_t=\sigma_a+\sigma_s$ where σ_a,σ_s are the absorption and total scattering coefficients respectively, the source term q and the stopping power S_{in} which reflects inelastic ("colour" shifting) interactions. The function $p(\mathbf{x},\mathbf{s}'\rightarrow\mathbf{s})$ is the normalized scattering phase which accounts for changes in the direction of the particle beam. Equation 14.1 contains all possible couplings of lowest order of parameter fields to the particle intensity $I(\mathbf{x},\mathbf{s};E)$.

The evaluation of the integro-differential equation 14.1 can be based on the formal integral solution

$$I(\mathbf{x},\mathbf{s};E)=I_s\cdot\exp[-\tau(R)]+\int_0^R dR'\cdot\exp[-(\tau(R)-\tau(R'))]\cdot Q(\mathbf{x}-R'\mathbf{s},\mathbf{s};E) \tag{14.2}$$

with the generalized source

$$Q(\mathbf{x},\mathbf{s};E)=q(\mathbf{x},\mathbf{s};E)+\sigma_s(\mathbf{x};E)\cdot\int d\omega'\cdot p(\mathbf{x},\mathbf{s}'\rightarrow\mathbf{s})\cdot I(\mathbf{x},\mathbf{s}';E) \tag{14.3}$$

$I_s=I_s(\mathbf{x}-R\mathbf{s},\mathbf{s};E)$ is the incident intensity (see figure 14.3 in the Appendix) and τ is the optical depth given by

$$\tau(R)=\int_0^R dR'\cdot\sigma_t(\mathbf{x}-R'\mathbf{s},\mathbf{s};E). \tag{14.4}$$

In equation 14.2 the term describing inelastic scattering is omitted. It will be separately considered in the next section.

Discretization methods for the evaluation of equation 14.2 are briefly discussed in the Appendix.

14.3 Mapping of data features onto visualization parameters

Important requests from the user's side on 3D visualization tools are convincing rules for the mapping of interesting data features onto the visualization parameters and "physically" meaningful actions of those parameters during the rendering process. Scientific visualization must produce well-defined pictures.

An advantage of the transport theory model proposed is that many different possibilities for synthesizing such a mapping suggest themselves. It can easily be demonstrated that all source-attenuation models in volume rendering (see e.g., [23, 18, 22, 24]) are covered by this model. A classification of possible mappings for the 3D scalar data field $F(\mathbf{x})$ assumed to be normalized into the range $[0,1]$ or derivates of it can be done as follows:

Source terms The term $q(\mathbf{x}, \mathbf{s}; E) \geq 0$ in equation 14.1 acts as an internal source for the particle intensity I. According to the spatial support q can be classified as point-like, line-like, surface-like, or volume-like.

Volume densities can be mapped onto a *volume source term* q_v in the form

$$q_v(\mathbf{x}, \mathbf{s}; E) = c(E) \cdot F(\mathbf{x}) \tag{14.5}$$

where the coefficient $c > 0$ describes a generic constant which accounts for the normalization of the intensity I according to equation 14.2. It may depend on the "colour" E. This mapping is only useful for the visualization of the spatial shape and decay of the data field. This approach was used to visualize the appearance of atmospheric data[24] and of the electron density of highly exited atoms[22]. The evaluation of equation 14.2 degenerates in the case of a pure volume source q_v into a summation of the field contributions along path 1 depicted in figure 14.3 in the Appendix. Disadvantages of this choice are the loss of enhanced depth information and a surpressing of detail. An example of the action of this visualization tool is given in plate 35. The appearance is similar to that of pictures from emission tomography or fluorescent materials.

To visualize isovalue surfaces of volume densities or strong discontinuities along surfaces in volume densities (e.g., [18, 24]) a *surface source term* q_s should be taken into account. It is given by

$$q_s(\mathbf{x}_s, \mathbf{s}; E) = c(E) \cdot (\mathbf{s} \cdot \mathbf{e}_s) \cdot \begin{cases} F(\mathbf{x}_s) & \text{forisosurfaces} \\ |F^+ - F^-|\,(\mathbf{x}_s) & \text{fordiscontinuities} \end{cases} \tag{14.6}$$

where $\mathbf{x}_s$ describes the coordinates of the surfaces, $\mathbf{e}_s$ is the local normal and $|F^+ - F^-|$ is the height of the discontinuity of the volume density perpendicular to the surface or the absolute value of the field gradient, respectively.

This method is especially popular in medical applications where an enhancement of boundaries between different tissue materials (see e.g., [18, 8]) is desired. An example for the appearance of an isovalue surface is given in plate 36, showing the role of the Lambertian factor $(\mathbf{s} \cdot \mathbf{e}_s)$ for the enhancement of depth formation. This mapping gives enhanced depth information and is also useful for the visualization of details (see also plates 37–40.

Point-like source terms q_P or *line-like source terms* q_L can be considered as special cases of the volume source term equation 14.12. Generally, "hot spots" in volume densities or interesting hyper-surfaces should be visualized with the mappings equations 14.5 or 14.6. The mapping, equation 14.6, is equivalent to the "diffuse" reflection term used in computer graphics which accounts for the description of light reflection from "very" rough surfaces, for example.

Absorption term The extinction term in the transport equation 14.1 causes an exponential attenuation of the intensity in equation 14.2 via the optical depth. Identifying the field or absolute value of the field gradient with σ_a in the form

$$\sigma_a(\mathbf{x}; E) = c(E) \cdot \begin{cases} F(\mathbf{x}) \\ |\text{grad } F(\mathbf{x})| \end{cases} \tag{14.7}$$

one gets visualization effects similar to x-ray pictures. An example for the influence of the absorption term on a non-zero initial intensity I_s is shown in plate 37. In addition, this picture also shows the attenuation of a surface source visualized with the mapping equation 14.6 and high-lightened by using a "specular" component in mapping equation 14.10.

Scattering terms Exploiting also the more "sophisticated" scattering term σ_s in the transport theory leads to more elaborate computation algorithms, e.g., the Monte Carlo method (see path 2 in figure 14.3 in the Appendix and equation 14.21).

This term should be incorporated in two different cases:

Selective enhancement of local fluctuations of the volume density can be modeled with a *volume scattering coefficient* σ_s^v by identifying

$$\sigma_s^v(\mathbf{x}) = c(E) \cdot F(\mathbf{x}) \tag{14.8}$$

and chosing an appropriate scattering phase function

$$p^v(\mathbf{x}, \mathbf{s}' \to \mathbf{s}) = c_f \cdot \delta(\mathbf{s}' \cdot \mathbf{s} - 1) + (1 - c_f) \cdot P_s(\mathbf{x}, \mathbf{s}' \cdot \mathbf{s}) \tag{14.9}$$

with $c_f \leq 1$. The first term accounts for forward scattering only and P_s is an arbitrary phenomeno-logical function such as that of Henyey-Greenstein[2]. This approach is suitable for the visualization of atmospheric data fields (clouds, dust, etc.)[2, 19].

A *surface scattering term* $\sigma_s^{\text{surf}}(\mathbf{x}_s)$ on a surface point $\mathbf{x}_x$ can be introduced to show two different effects:

An enhanced visualization of isovalue surfaces or of sharp boundaries between volume regions having different densities (e.g., in medical applications) can be obtained by introducing a *specular scattering term*

$$P^{\text{spec}}(\mathbf{s}' \to \mathbf{s}) = c_{\text{spec}} \cdot \delta(\mathbf{s} - \mathbf{s}_{\text{spec}}). \tag{14.10}$$

The phase function P^{spec} defines the shininess of the surface if a Phong-like smoothing of the δ-function around the specular direction $\mathbf{s}_{\text{spec}}$ is chosen. The location of the specular reflecting parts on the surface can be artifically chosen by introducing additional external sources. The role of the specular reflection for the enhancement of the depth information is demonstrated in plates 37–40.

A combination of a transmitting and a backscattering phase function

$$P^s(\mathbf{s}' \to \mathbf{s}) = c_f \cdot \delta(\mathbf{s}' \cdot \mathbf{s} - 1) + c_b \cdot \delta(\mathbf{s}' + \mathbf{s}), \quad c_f + c_b = 1 \tag{14.11}$$

defines the transparency of the surface depending on the relation of the forward and backward components c_f and c_b.

Plate 39 demonstrates the combined mapping of the features of a volume density (energy density of a vibrating crystal lattice) onto the volume source term equation 14.5, the surface term equation 14.6, and the specular component equation 14.10. The visualization of the iso-surfaces underlines in this example the spatial decay of the interatomic potential.

Almost all volume rendering tools use a combination of the mappings equations 14.5, 14.6, 14.9–14.11. Plate 41 is a visualization of a medical CT-data set showing the effect of this combination.

Colour shifting term In visualization applications data sets very often appear representing field densities with varying sign, e.g., charged fields or fields given relatively to the mean such as pressure or temperature. In these cases the colouration is an essential tool (see e.g., [6]).

In general, all parameter fields $C(\mathbf{x}; E)$ in equation 14.1 depend on the space coordinates $\mathbf{x}$ and on the energy parameter E ("colour"). Using the decomposition

$$C(\mathbf{x}; E) = C_x(\mathbf{x}) \cdot C_E(E) \tag{14.12}$$

one gets two parallel mapping rules for data field features. Plates 35–39 and 41 have been rendered with this simple method to enhance the different appearance of volume and surface sources.

Another approach to use colour for feature enhancement can be introduced by incorporating the more "physical" stopping power term in equation 14.1 which represents inelastic scattering processes. The influence of the stopping power term can be taken into account by the shift

$$E \rightarrow E - \int_0^R S_{in}(\mathbf{x} - R'\mathbf{s})dR' \tag{14.13}$$

to be inserted in all expressions. This term generates for the discrete colour values a scaling of the form

$$I_i(s + \Delta s) = I_i(s) \cdot [1 - f_i \cdot S_{in}(s) \cdot \Delta s], \quad i = R, G, B \tag{14.14}$$

to be inserted into the recursion relations equations 14.23 or 14.24 in the Appendix. The constant factors fi represent appropriate scalings.

Identifying $S_{in}(\mathbf{x})$ with the data field the energy ("colour") will be shifted up or down locally, depending on the sign of the field value. Plate 40 shows a visualization example for such 3D fields by using equuation 14.14 with $f_i = \text{const} \cdot (-1, 0, 1)$ in addition to volume and surface source terms.

14.4 Tools for enhancement of critical features

All of these tools for visualization of 3D scalar data fields are "classic" ones when considered under the aspect of the transport theory model. Recognizing the applications of the transport theory in so many areas additional tools for enhancing special (hidden) features of data fields might be overtaken in an appropriate form.

Role of texture – "quality" of data An essential part of the interpretation process in science is the comparison of related data fields. For instance, models for the representation of complementary data fields use the composition of visualization tools as surface topography, colouration, and transparency effects[21].

Another important class of interpretation problems is concerned with the comparison of data sets describing the same effect but resulting from different methods such as closed-form analytic solution, numerical simulation, measurement, etc. An example of data sets originating from an analytical model and a corresponding Monte Carlo simulation is shown in figure 14.1.

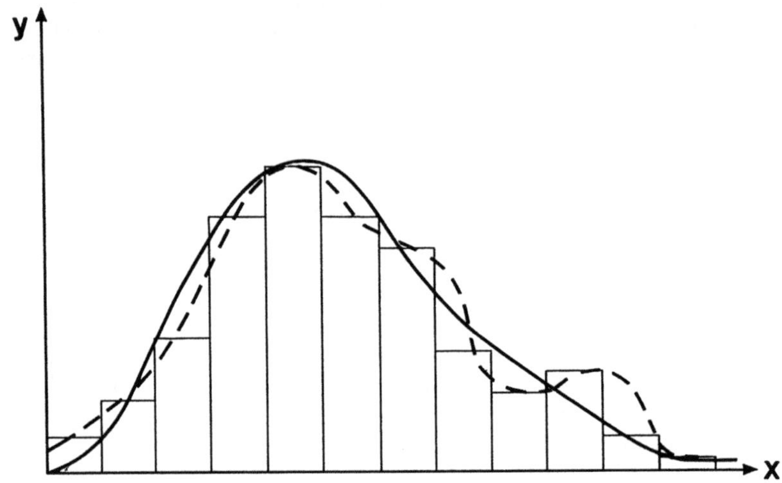

FIGURE 14.1. of analytic model data (——) with results from a Monte Carlo Simulation (- - -) (smoothed histogram)

For visualizing an enhanced appearance of such differences of related data fields the model of random texturing for representing macroscopic deviations of volume densities or surface heights[17] seems to be appropriate.

The data field $F_2(\mathbf{x})$ to be compared with the field $F_1(\mathbf{x})$ can be decomposed as

$$F_2(\mathbf{x}) = F_1(\mathbf{x}) + \Delta F(\mathbf{x}) \tag{14.15}$$

where ΔF represents, for instance, imperfections of the measurements or calculations. The spatial average of ΔF represents systematic deviations and will not be equal to zero generally. The deviation field $\Delta F(\mathbf{x})$ can be described by its statistical parameters mean, variance, and autocorrelation $C(\Delta\mathbf{x})$. It is convenient to take a Gaussian statistics with

$$C(\Delta\mathbf{x}) = \frac{\langle \Delta F(\mathbf{x}+\Delta\mathbf{x}) \cdot \Delta F(\mathbf{x}) \rangle}{\langle \Delta F(\mathbf{x}) \rangle^2} = \exp\left[-\frac{\Delta\mathbf{x}^2}{2C_\tau{}^2}\right] \tag{14.16}$$

where the brackets denote spatial averaging and the autocorrelation length C_τ is assumed to be of order of the grid length.

These statistical parameters are mapped onto corresponding parameters of the particle intensity I via the linear transport equation. Equation 14.2 generates the propagation of field deviations along the particle ray in the form

$$I_2(\mathbf{x}, \mathbf{s}; E) = H \cdot I_1(\mathbf{x}, \mathbf{s}; E) + N \tag{14.17}$$

where H and N depend on the deviations of the parameters of the transport theory according to the mappings chosen.

The role of random texture for the comparison of data fields on hyper-surfaces can easily be demonstrated. Assuming the interesting surface will be visualized by a surface source term 14.6 only the variance of the particle intensity is proportional to the variance of the data field deviations .

$$\sigma_I^2 = \frac{\langle (I - \langle I \rangle)^2 \rangle}{\langle I \rangle^2} = \text{const.} \frac{(F_2 - F_1)^2}{F_1^2} \tag{14.18}$$

where F_1, F_2 are the data sets on the surface to be compared. The autocorrelation of I has the same form as equation 14.14 with appropriately transformed correlation lengths. The constant in equation 14.18 should be chosen such that ${\sigma_I}^2$ varies from zero to values larger than 10. Then the natural texturing model[17] can be applied where ${\sigma_I}^2$ generates more or less strong non-Gaussian intensity fluctuations on the screen. The autocorrelation length influences the size of the random texture patterns such that the granularity depicts the fineness of the underlying mesh space. Examples for this method are visualized for a 2D distribution of ions implanted in a semiconductor[10]. Results from Monte Carlo simulations are compared with those of analytic models according to figure 14.1. In plate 42 the texture shows strong fluctuations at the tails of the distribution typically for the noise content of Monte Carlo simulations.

Considering a mapping of the data field onto the volume source term according to 14.5 equation 14.2 shows the dependence of I on the spatial average of the source deviations $\Delta q(\mathbf{x})(\approx \Delta F(\mathbf{x}))$ along the ray path. Assuming the data field $F_2(\mathbf{x})$ only deviates significantly on a few grid points from $F_1(x)$ the intensity variance is given by

$$\sigma_I^2 = \text{const.} \frac{\left(\sum_{i=1}^{k} \Delta q(\mathbf{x}_i) \Delta s \right)^2}{\left(\int_0^R dR' \cdot q(\mathbf{x} - R'\mathbf{s}, \mathbf{s}; E) \right)^2} \tag{14.19}$$

where k counts the deviation points along the path. As long as k is small compared to the number of grid points along the path σ_I^2 will be different from zero. The texture pattern indicating some strong point-like data field deviations is shown in plate 43.

The form equation 14.15 of the disturbed intensity I_2 at the screen also suggests the incorporation of filtering methods into the rendering process to enhance the visual appearance of noisy data fields.

Visualization of symmetry effects – channeling In practice, many data fields have to be checked for (hidden) symmetry properties. The effect of channeling of charged particles in crystalline solids[20, 1] can be used to indicate field symmetries by specific patterns appearing on the surface of the data field. This is an example for visualizing data field features by "virtual" particles other than "photons".

Assigning to the test particle an "electric" interaction with the field data than it will be scattered by the field and follow the path 2 depicted in plate 35 in the Appendix. If there is any symmetry in the data field, e.g., symmetric "strong" scattering centers, then specific directions along "symmetry channels" will be preferred (see figure 14.2).

Assuming a binary collision model and Rutherford scattering, for example, [20] the path of the "charged" test particle through the data field can easily be followed.

Letting such test particles cross the data field in a form of a pencil beam, a specific pattern will appear on the screen. Plate 44 shows the "classic" rendering of a volume field and in addition the scattering pattern of 1000 "charged" test particles caused by

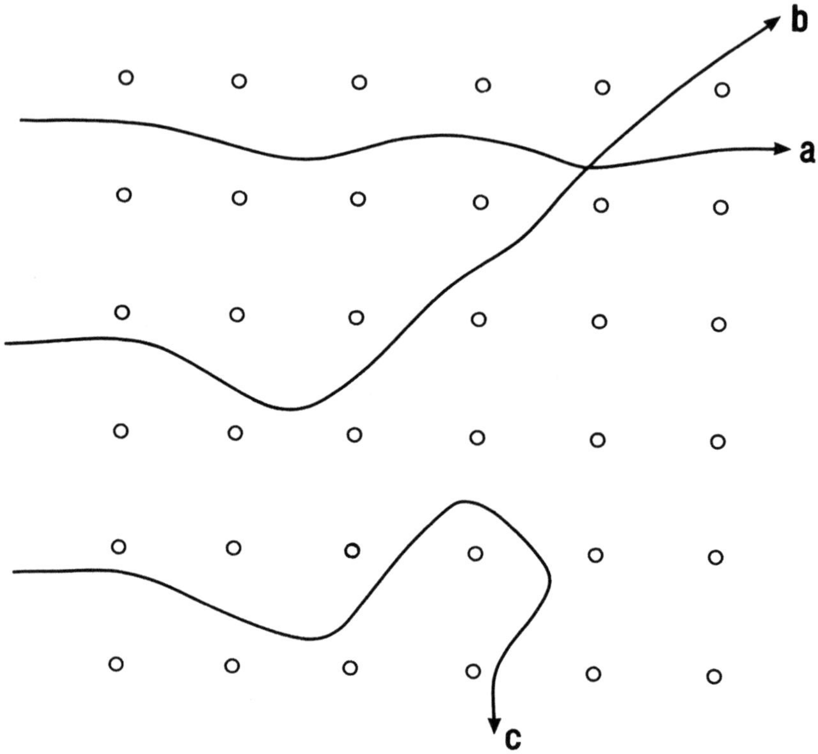

FIGURE 14.2. Movement of charged test particles in a data field, (a) and (b) channeling along different symmetry axes, (c) randomized movement

diamond-like symmetry properties of the data field. The advantage of this method is that certain channeling effects occur even in the case of a disturbed symmetry.

14.5 Appendix: evaluation of the transport equation

The evaluation of the integral form equation 14.2 of the linear transport theory can be done by discretization. This procedure describes a piecewise linear transfer of "virtual" particles, incident with an intensity I_s along the volume surface on the right, through the volume considered (see figure 14.3).

Equation 14.2 has the usual form of a linear operator equation

$$I = I_0 + K * I \tag{14.20}$$

such that a solution can be obtained by expanding I into a Neumann series

$$I = I_0 + \sum_{i=1}^{N} (K*)^i I_0 \tag{14.21}$$

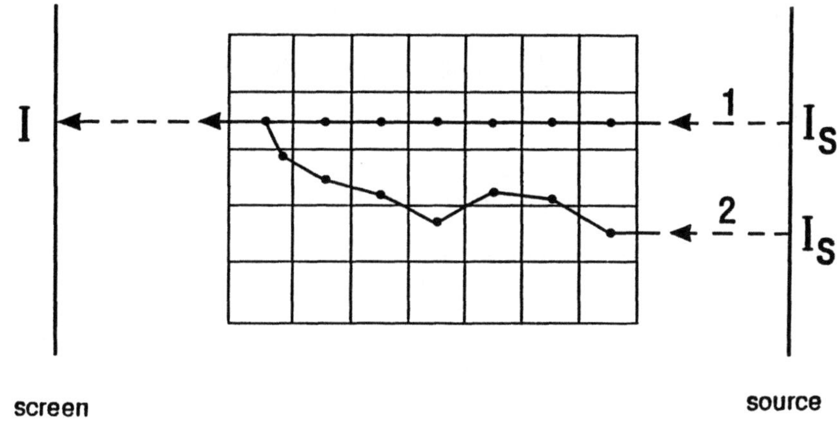

FIGURE 14.3. Possible simulated particle paths

where the integral operator K is given by

$$K = \int_0^R dR' \cdot \exp\left[-(\tau(R) - \tau(R'))\right] \cdot \sigma_s(\mathbf{x} - R'\mathbf{s}; E) \cdot \int d\omega' \cdot p(\mathbf{x}, \mathbf{s}' \rightarrow \mathbf{s}). \quad (14.22)$$

I_0 is given by equation 14.2 by neglecting the scattering term. The value of I_0 at the screen can be obtained by following the line-of-sight path 1 in the figure and using the simple recursion rule

$$I_0(s + \Delta s) = I_0(s) \cdot \exp[-\sigma_t(s + \Delta s) \cdot \Delta s] + q(s + \Delta s) \cdot \Delta s. \quad (14.23)$$

This formula has been widely used in source-attenuation models in volume rendering by mapping the data field onto σ_t and 9 (see e.g.,[18], 21]). The exponential attenuation factor has been often approximated by the linear expansion term ("opacity").

"Smoother" results can be obtained (see e.g.,[3]) by using the trapezoid rule for the integration in equation 14.2 giving

$$I_0(s+\Delta s) = [I_0(s) + .5 \cdot q(s) \cdot \Delta s] \cdot \exp[-.5 \cdot (\sigma_t(s+\Delta s) + \sigma_t(s)) \cdot \Delta s] \quad + .5 \cdot q(s+\Delta s) \cdot \Delta s. \quad (14.24)$$

The values of the local scattering parameter fields σ_t, σ_s, and q at the endpoints of the path elements can be evaluated by interpolating about the eight values at the surrounding lattice points taking a distance dependent weighting into account.

Generally, the choice of the length of the path elements is a delicate task. For very inhomogeneous data fields Δs has to be of the order of the diameter of the finest details. For relatively homogeneous parts of the data field a speed up of the calculations can be obtained by using

$$\Delta s = -\Delta \overline{s} \cdot \ln \text{ (ran)} \quad (14.25)$$

where $\Delta \overline{s}$ is a suitable chosen mean value and ran is a uniformly distributed random number. For the test pictures a mean path length $\Delta \overline{s} = 1/2 \cdot$(grid length) was chosen. The volume rendering via equation 14.23 or 14.24 is very time consuming. Special sampling methods as discussed in[5, 18] can be very helpful.

The additional terms I_i in equation 14.21, describing multiple scattering events (see path 2 in the figure), can be evaluated by the formal recursion relation

$$I_i(\mathbf{x}, \mathbf{s}; E) = \int dR \cdot \exp[-(\tau(R) - \tau(R'))] \cdot \overline{Q}_{i-1}(R') \qquad (14.26)$$

where $\overline{Q}_{i-1}$ is given by

$$\overline{Q}_{i-1}(R') = \sigma_s(\mathbf{x} - R'\mathbf{s}) \cdot \int d\omega' p(\mathbf{x}, \mathbf{s}' \to \mathbf{s}) \cdot I_{i-1}(\mathbf{x} - R'\mathbf{s}, \mathbf{s}'; E) \qquad (14.27)$$

The evaluation of this general formula should be done for non-trivial scattering phases equation 14.9 with the aid of the Monte Carlo method[15, 7].

In practice, the tracing of the particles through the volume will be done from the view-point to the back. This is equivalent to the evaluation of the adjoint transport equation which follows from equation 14.1 by reversing the sign of s.

14.6 References

[1] E. Uggerhoj A. H. Sorensen. The Channeling of Electrons and Positrons. *Scientific American*, pages 70–77, June 1989.

[2] J.F. Blinn. Light Reflection Functions for Simulation of Clouds and Dusty Surfaces. *Computer Graphics*, 16:21–29, 1982.

[3] M. Keeler C. Upson. VBUFFER: Visible Volume Rendering. *Computer Graphics*, 22(4):59–64, 1988.

[4] S. Chandrasekhar. *Radiative Transfer*. Dover, 1960.

[5] R.L. Cook. Stochastic Sampling in Computer Graphics. *ACM Transactions on Graphics*, 5(1):51–72, January 1986.

[6] A.J. Olson D.S. Goodsell, S. Mian. Rendering of Volumetric Data in Molecular Systems. *Journal of Molecular Graphics*, 7(1):41–47, March 1989.

[7] G.I. Marchuk et al. *The Monte Carlo Methods in Atmospheric Optics*. Springer Verlag, Berlin, 1980.

[8] K. Tiede et al. Investigation of Medical 3D-Rendering Algorithms. *IEEE Computer Graphics and Applications*, pages 41–53, March 1990.

[9] St. Pizer H. Fuchs, M. Levoy. Interactive Visualization of 3D Medical Data. *Computer*, 22(8):46–51, 1989.

[10] W. Krueger H. Ryssel, J. Lorenz. Ion Implantation into Non-planar Targets: Monte Carlo Simulations and Analytic Models. *Nucl. Instr. and Meth. B 19/20*, pages 45–49, 1987.

[11] K.E. Torrance H.E. Rushmeier. Extending the Radiosity Method to Include Specularly Reflecting and Translucent Materials. *ACM Transaction on Graphics*, 9(1):1–27, June 1990.

[12] K.J. Hussey. Image Processing as a Tool for Physical Science Data Visualization, Course Notes Number 28. Technical report, ACM SIGGRAPH, 1987.

[13] W.R. Martin J.J. Duderstadt. *Transport Theory*. Wiley, 1979.

[14] B.P. Van Herzen J.T. Kajiya. Ray Tracing Volume Densities. *Computer Graphics*, 18(3):165–174, 1984.

[15] J.T. Kajiya. The Rendering Equation. *Computer Graphics*, 20(4):143–150, 1986.

[16] P.F. Zweifel K.M. Case. *Linear Transport Theory*. Addison-Wesley, 1967.

[17] W. Krueger. Intensity Fluctuations and Natural Texturing. *Computer Graphics*, 22(4):213–220, 1988.

[18] M. Levoy. Display of Surfaces from Volume Data. *IEEE Transactions on Computer Graphics and Applications*, pages 29–37, May 1988.

[19] N. Max. Light Diffusion through Clouds and Haze. *Computer Vision, Graphics and Image Processing*, 33:280–292, 1986.

[20] O.S. Oen M.T. Robinson. Computer Studies of the Slowing Down of Energetic Atoms in Crystals. *Physical Review*, 132:2385–2398, December 1963.

[21] J.F. O'Callaghan P.K. Robertson. The Application of Scene Synthesis Techniques to the Display of Multidimensional Image Data. *ACM Transactions on Graphics*, 4(4):247–275, 1985.

[22] et al. Ruder, H. 'Line-of-Sight Integration: A Powerful Tool for Visualization of Three-Dimensional Scalar Fields. *Computers & Graphics*, 13(2):223–228, 1989.

[23] L. Hesselink S. Jaffey, K. Dutta. Digital Reconstruction Methods for Three-Dimensional Image Visualization. *Proceedings of the Society of Photo-Optics Instrumentation Engineering*, 507:155, 1984.

[24] P. Sabella. A Rendering Algorithm for Visualizing 3D Scalar Fields. *Computer Graphics*, 22(4):51–58, 1988.

15 Visualization of 3D Empirical Data: The Voxel Processor

W. Huiskamp, A. A. J. Langenkamp, and P. L. J. van Lieshout

ABSTRACT
Image processing is an area where the application of parallelism is very suitable, because of the large amounts of data involved. In this particular case, 3D images (voxel-images) have to be processed interactively. Operations on the voxel-images involve 3D image processing and visualization from arbitrary angles. The paper describes the development of a prototype voxel-processor, based on a network of T800 transputers.

15.1 Introduction

The Physics and Electronics Laboratory TNO (TNO-FEL) in the Hague is a part of the TNO Division of National Defence Research (HDO-TNO). The activities of TNO-FEL focus primarily on operational research, information processing, communication and sensor systems. To support the fast data-processing usually required in sensor systems applications, research was started into parallel processing. This research has now resulted in two other major application areas : real-time computer generated imagery and 3D image analysis, processing and visualization. With the growing availability of 3D scanning devices, the need for high performance processing and display systems increased enormously. This paper describes the development of an experimental parallel processing system for the visualization of three dimensional voxel-images. The aim is the visualization of the (unknown) object in such a way that its spatial structure can be understood. An additional demand is that the system is fast enough to be used interactively. Because of the large number of voxels involved, a considerable processing capacity is required. Processing the data in parallel on a network of Transputers provides the necessary computing power. The volume data may be visualized in several ways, involving operations like object transformation, hidden-surface removal, depth-shading and cross-sectioning. The main advantages of the developed system over dedicated hardware solutions are :

- Speed
- Flexibility
- Expandability
- Ease of design

15.2 The voxel data

Volume images are normally represented as a series of parallel two dimensional slices. These slices may have been obtained from several possible sensor systems, examples are Computer Tomographic- (CT), Nuclear Magnetic Resonance- (NMR), Ultra Sounding or Optical- (LASER) scanners.

Voxel representations are very suitable for applications with 3D empirical data. However synthetic data may also be used, for example solid modelling or fluid dynamics simulations. Before volume rendering became feasible, experts had to interpret the slices to deduce the

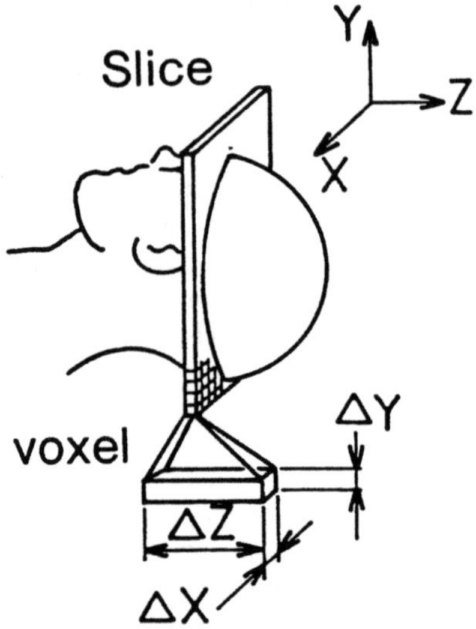

FIGURE 15.1. Voxel Definition

3D information. Until recently, computer assisted techniques to visualize the volumes were based on displaying contours only, because of the processing time involved. These contours often had to be traced manually from the actual data. Full use of the 3D data could only be made through off-line computing. Several architectures based on dedicated hardware have been proposed to increase performance[1, 4]. Such a dedicated system however has the disadvantage of inflexibility to any change in rendering options or object sizes (also the cost is high). This explains the reason for TNO-FEL to apply a system of programmable (low cost) processors operating in parallel.

During development of this prototype system, object data was obtained from an experimental Confocal LASER Scanning Microscope (CLSM). The CLSM can be focussed on several consecutive layers of the object, producing a slice of data for each layer. A slice typically consists of 256*256 volume elements (voxels), with an intensity resolution of 8 bits per voxel (figure 15.1. The number of layers may vary, but a typical value is 32 to 256 (figure 15.2). This data structure is called a Voxel-image. Examples of application areas for the CLSM are medical- and biological-research and inspection of Integrated Circuits. Voxel-data sizes depend largely on the sensor type, in CT scans for example it is possible to get resolutions of 512*512*128 with 12 bits per voxel. The developed voxel processor has the modularity to deal with varying size- and performance- demands.

15.3 The 3D reconstruction

The Voxel data consists of a block in 3D space (figure 15.3). Displaying this data under different angles on a 2D screen involves a 3D transformation of the object space to the display space (figure 15.4). Basically such a transformation consists of a vector-matrix multiplication on each voxel coordinate (i.e., a vector). These matrices for a rotation about

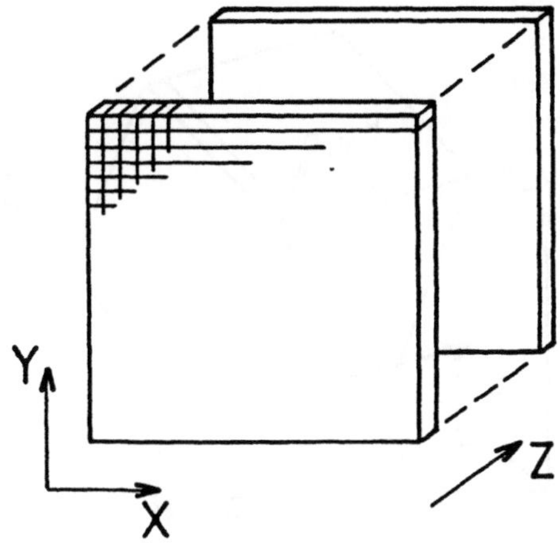

FIGURE 15.2. Voxel Image

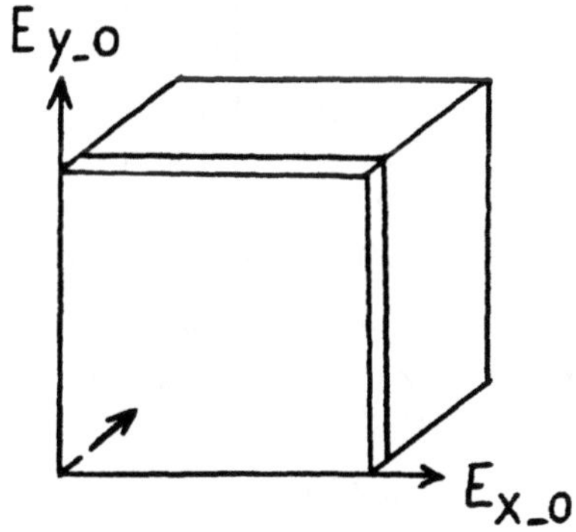

FIGURE 15.3. Object Space

the X-, Y-, and Z- axis are combined into a single matrix before the actual multiplication:

$$R = Rz * Ry * Rx = \begin{vmatrix} (cB \cdot cC) & (sA \cdot sB \cdot cC - cB \cdot sC) & (cA \cdot sB \cdot cC + sA \cdot sC) \\ (cB \cdot sC) & (sA \cdot sB \cdot sC + cB \cdot cC) & (cA \cdot sB \cdot sC - sA \cdot cC) \\ -sB & (sA \cdot cB) & (cA \cdot cB) \end{vmatrix}$$

('c' for cos, 's' for sin).

Since vector-matrix multiplication is a linear operation and all coordinates have to be transformed, it is not necessary to perform this multiplication for each coordinate. We may instead use three simple additions to step from one transformed coordinate to the next. This method offers a considerable reduction in the computational load.

The first step is to transform the unit-vectors from object- space into display-space, by multiplication with the previously calculated rotation matrix R.

$$\begin{vmatrix} nx.x' \\ nx.y' \\ nx.z' \end{vmatrix} = \begin{vmatrix} 1 \\ 0 \\ 0 \end{vmatrix} \cdot R$$

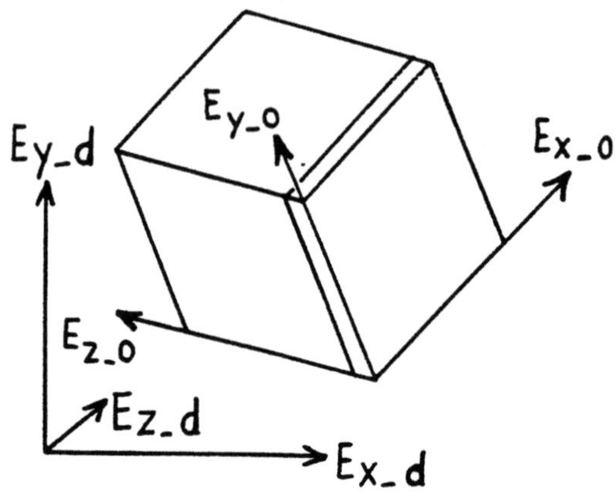

FIGURE 15.4. Display Space

$$\begin{vmatrix} ny.x' \\ ny.y' \\ ny.z' \end{vmatrix} = \begin{vmatrix} 0 \\ 1 \\ 0 \end{vmatrix} \cdot R$$

$$\begin{vmatrix} nz.x' \\ nz.y' \\ nz.z' \end{vmatrix} = \begin{vmatrix} 0 \\ 0 \\ 1 \end{vmatrix} \cdot R$$

The transformed coordinates (x', y', z') of voxel (x, y, z) are now found with :

$$\begin{vmatrix} x' \\ y' \\ z' \end{vmatrix} = x^* \begin{vmatrix} nx.x' \\ nx.y' \\ nx.z' \end{vmatrix} + y^* \begin{vmatrix} ny.x' \\ ny.y' \\ ny.z' \end{vmatrix} + z^* \begin{vmatrix} nz.x' \\ nz.y' \\ nz.z' \end{vmatrix}$$

The projection of the 3D data onto a 2D surface (the screen) involves the hidden surface elimination : 'distant' voxels are obscured by 'closer' voxels if they are projected on the same location on the screen. Comparing the z-value of a new pixel with the z-value of the pixel already prèsent on that screen location (Z-buffer algorithm), is avoided by traversing the voxel-data in a back-to-front direction. When generating the screen this way, new pixels can simply overwrite any old value (Painters algorithm). Several ways of rendering the transformed data on the screen are possible, the currently implemented options are :

- Display the object's intensity, as seen from the selected orientation ('front view'). (plate 45)
- Display the object's 'distance' from the screen at each location, resulting in a realistic depth illusion. ('depth shading').
- Display the object's density at each screen location, ('integrate function').
- Display an intensity related to the layer from which the visible voxel originated. ('layer view'). (plate 46)
- Select a 'Volume-Of-Interest' within the available voxel data (this volume must be block-shaped). Through this option uninteresting or disturbing parts of the voxel-image may be 'peeled away' (plate 48).

- Select a cutting plane through the object; voxels in front of this plane will not be visualised. This option will create a cross-section through the object after rotation. (plate 47)
- Select a threshold; voxels with a value below this threshold will become transparent.
- Edit and select different colour look-up tables. This feature enables the use of pseudo-colours or grey-scale transforms for certain intensity values, thereby increasing the visibility of interesting areas.

The original images from the scanning device tend to be noisy in many cases, so noise filters are needed. Further image analysis operations (edge detectors etc.) are also provided. Currently implemented 3D image processing algorithms are :

- Mean filter
- Sobel and Roberts edge detectors
- Laplace filter
- Median filter.

These filters are based on their 2D counterparts and operate on a (3*3*3) space. Computation of histogram information over (part of) the voxel- data is also implemented. This data may be plotted or used by specific image processing functions like automatic thresholding, histogram equalization or edge detection.

15.4 Parallel processing

Computer applications tend to need increasing amounts of processing capacities. Single processor systems are reaching the limits of performance improvements. It is obvious that using more processors running in parallel should provide (theoretically) unlimited power. Most existing multi-processor systems use a common communications channel (the bus) for interconnections. With a growing number of processors the bus capacity becomes a bottle-neck for system performance. Communication bandwidth of the network must be increased also when processors are added. Providing processors with direct connections for all data exchange will supply this increased bandwidth.

Several classes of multi-processor systems may be defined :

- Single Instruction Multiple Data (SIMD). Each processor in the network will execute the same instruction (synchronously) on different data. Array processors fall in this class. Examples are image processing applications where each processor performs the same filter operation on a different part of the image.
- Multiple Instruction Multiple Data (MIMD). Processors can all be running different programs, possibly sending results to others when they are finished. Examples are pipelined systems or multi-user applications.

Many existing sequential programs could benefit from being able to perform more than one action at a time. It is however generally not trivial to implement a parallel program on a processor network. Problems arising are :

- Decomposing the problem in a number of processes running in parallel

- Allocate processes to processors and select the network topology
- Load-balancing the processors
- Distributing data across the processors
- Efficient inter-processor communication
- Synchronization between processors
- Debugging the software.

The INMOS T800 Transputer is a computer-on-a-chip, containing a 32 bit RISC ALU, a 64 bit Floating-Point Unit, memory and four high-speed (1.5 MByte/s) input/output links for point-to-point communication[3] (see figure 15.8). The Transputer was specifically designed for efficient parallel processing: it is a high performance component (10 MIPS, 1.5 MFLOPS), with an on-chip process scheduler and low-overhead communication facilities. A network of Transputers may be constructed by connecting them via links. Each Transputer in a network has (private) local memory to store program and data. Transputers may be programmed in high level languages like Pascal, Fortran or C. These languages must have facilities added to implement the special features of the Transputer (processes running in parallel, communication etc.). OCCAM is a language that was created by INMOS to describe parallel processing and communication via channels[2]. In fact the Transputer may be considered a hardware implementation of OCCAM. Transputer versions without the floating-point unit are also available : the T414 (32 bit) and the T212 (16 bit).

Transputer networks belong to the MIMD class of parallel processing systems, all nodes in a network are basically independent units, communicating and synchronising only when necessary. A MIMD network is the most flexible solution to parallel processing, since part of the network may actually be operating as SIMD. At TNO-FEL, research has concentrated on the Transputer as the computational element in parallel processing applications, because of its useful features, high performance and software support.

Parallelism may be accomplished in 3D image processing by splitting up the computations in display-space or in data-space:

- Display space parallelism implies that each processor is assigned to a certain area of the resulting image (e.g., a number of scanlines). Since views of the rotated voxel-image will be generated, this solution means that each processor must have access to the complete voxel-image. Complete access is possible when a voxel-image copy is stored in each processor (large memory requirement) or alternatively, processors could request voxel-data elements when needed from a central store (communication overhead). Load-balancing may be a problem, since the most computation intensive parts in the display-space will shift according to the rotation angle. Ray-tracing is a typical example where parallel processing in display space is often used. The load-balancing problem can be tackled by implementing a processor farm. In this construction a controller process "farms out" a new piece of work (i.e., a part of the display) to each processor in the network as soon as this has finished work on a previous part. The controller does not need to know which processor will actually perform the job.

- Data space parallelism is based on access of a limited part of the original voxel-image. This implies that each node is assigned to a section of the voxel-image, which

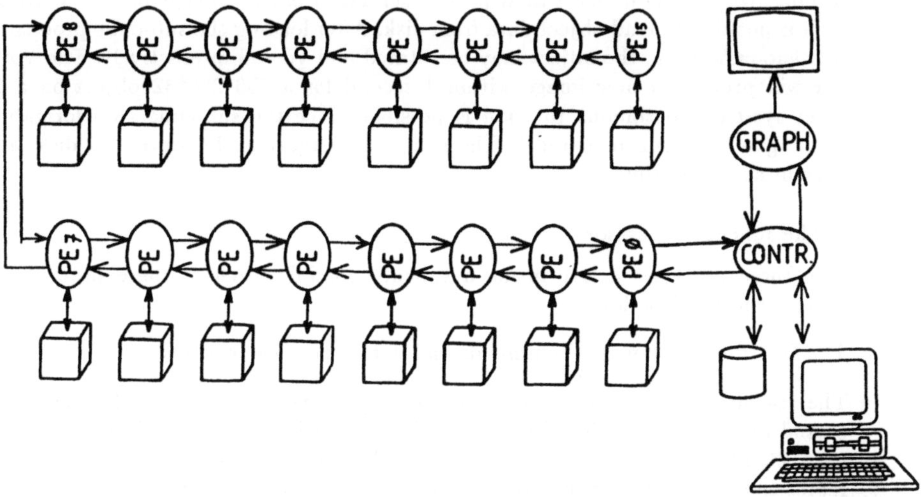

FIGURE 15.5. System Architecture

is stored locally. A node will produce the contribution of the local data to the result. The actual result will be available after combining (merging) all the contributions. The advantages of this method over the previous one are :

- Less memory requirement
- Fast access to the (local) voxel-data
- Good load-balancing, all contributions will need the same computation time, when the voxel-data sizes are equal Disadvantages are : the overhead of the merging operation and the fact that some data calculated by the nodes may not be needed in the final result.

Typically, a large number of views will be generated from a single (large) dataset. Therefore, the lower communication need of the second method was the reason to choose data-space parallellism in the voxel-processor system. Each Transputer holds a data segment according to its position in the network. A segment consists of a number of 2D slices.

15.5 System architecture

Figure 15.5 shows a schematic representation of the voxel processor architecture. Ellipses are used to represent modules running in parallel. Parallelism was achieved in several ways, the most important step is dividing the object data into a number of (equally sized) sub-cubes, where each sub-cube has been assigned to one Transputer. The modules will be explained in more detail below.

15.5.1 The controller

The operator communicates with the 'user-interface' part, which interprets and executes commands. The controller process is capable of sending commands to a Transputer based framegrabber, which may be used to acquire voxel data from some type of sensor. Alternatively, it is possible to read object data from disk. After loading the sub-cube processors with their sub-cube data, the object may be rotated and viewed interactively. The Voxel processor will produce a new image within 1 second for a 256*256*32 object on a 16 Transputer system. Continuous rotation is possible, since the sub-cube transformation and the merging may run in parallel. The resulting images can be stored on disk, and read in again at a later time.

15.5.2 The subcube processor

There are three processes active inside a subcube module, each of these processes is assigned to a specific function :

- The Distributor, transfers commands and data across the network.

- The transformation process performs the object transformation and generates a partial result for the assigned sub-cube. The dimensions of the resulting image will be 512*512 pixels for an object of 256*256*32 voxels. A sub-cube partial result represents the intensity value for each pixel on the screen, except for the 'integrate' function, in which case a voxel count will be produced. The sub-cube data (the voxels) are loaded only once for each new object and will not be changed during the transformations. The Transputer will begin processing its data after receiving a command, which includes the transformed unit vectors and the selected type of rendering (e.g., front view, depth shade etc.). Besides performing the voxel-image transformation, this module is also used for the 3D image processing operations. For this end, memory is reserved to store both the original voxel-image and a filtered version.

- The merger, combines partial results to get the complete resulting image. This result is transferred to the controller process where it will be stored and displayed. The merger process receives 2D partial results from the local sub-cube processor and from its direct neighbour. In order to combine the two partial results in a correct way, the merger needs some additional data : the subcube's priority. The priority is based on the z-value of the subcube's transformed origin. The lowest priority is given to the sub-cube with the largest distance from the viewer. The partial results of this subcube will be obscured by any sub-cube result of a higher priority (figure 15.7 and 15.8). The merging operation is mainly performed with the 2D block-move instruction (DRAW2D) of the T800 Transputer, which can be used when the sub-cube results are mapped on top of each other in order of increasing priority. The DRAW2D can not be used for the 'integrate' function, in which case the numbers in the sub-cube results corresponding to the same pixel have to be added.

15.5.3 The graphics subsystem

This unit is used for controlling a framebuffer in order to display the resulting images. A second function of this board is the acquisition of slice-data. The graphics board is in fact a framegrabber, capable of digitizing analogue video. Each recorded slice is the result of a number of (digitally) integrated video frames. This integration is used to reduce the noise level. The voxel data-set will be temporarily stored in the processor nodes, after which it may viewed first before transferring the data to disk.

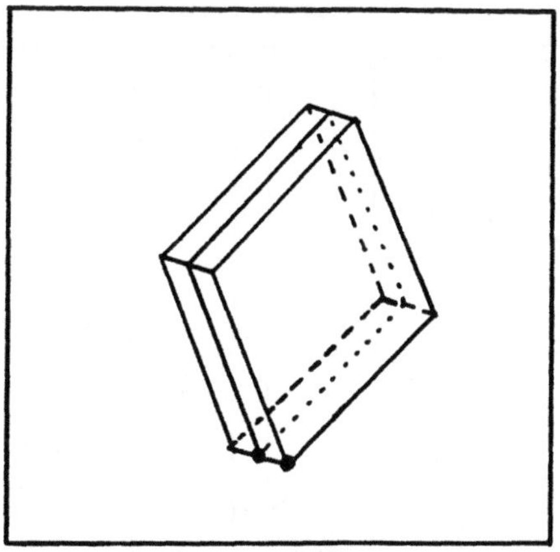

FIGURE 15.6. Subcube Offset

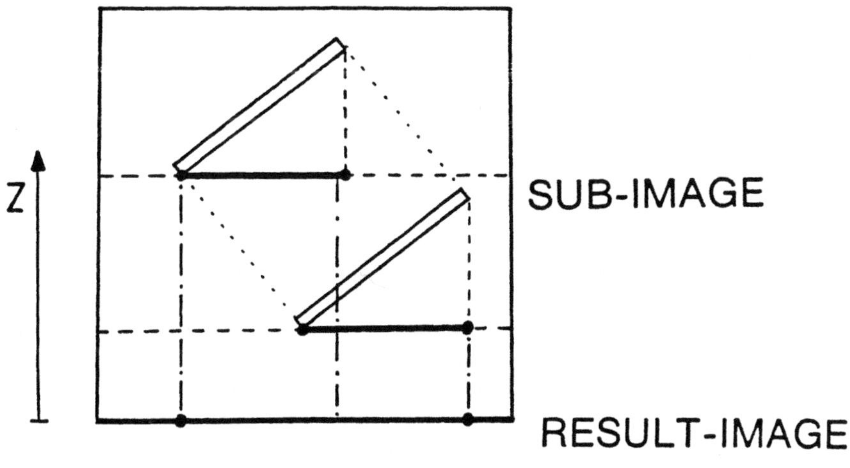

FIGURE 15.7. Subcube Priorities

15.6 Implementation remarks

- The Voxel processor is built up entirely with of-the-shelf hardware. Each processor board offers two T800's with 2 MByte of memory each. Other boards used, are the Display System with an on-board framegrabber, a T800 and 1 MByte of video-ram. Physically the system consists of a 19" cabinet with 10 single euro-sized boards installed. The host system in the Voxel- processor is an IBM-AT. All the program code was written in OCCAM.

- A communication layer was integrated into the system. This layer provides data and command transport to all processes, and it is also capable of sending (debug) messages from each process to the operator screen. Whenever possible, special processes were assigned to communication in order to achieve maximum efficiency of the Transputer Links.

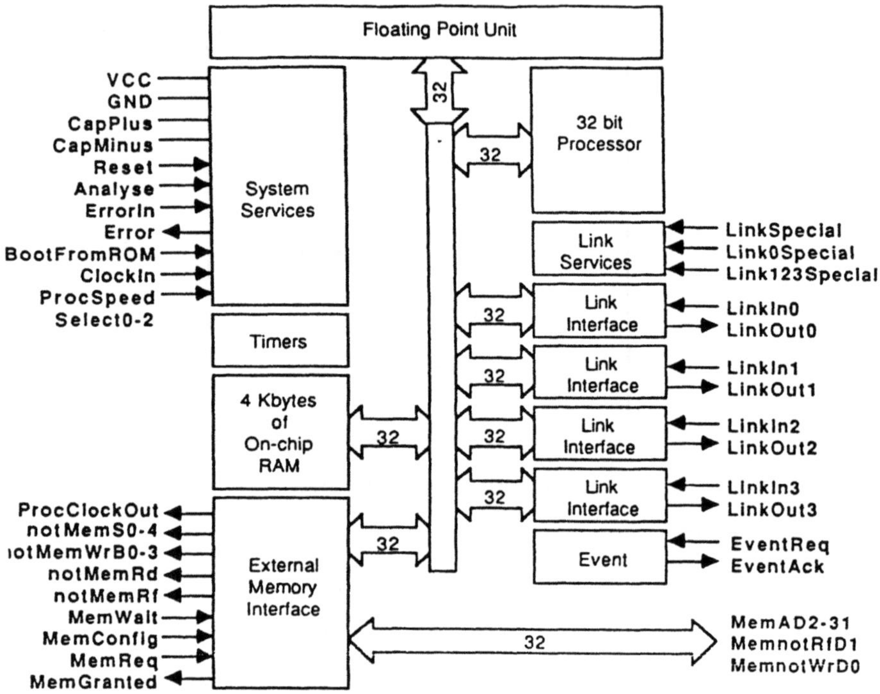

FIGURE 15.8. T800 Block Diagram

- The system is very flexible in the dimensions of the objects that are to be transformed. Basically the only limitation is the memory size of each Transputer (currently 2 MByte). These dimensions (and a few derived values) are declared in a library. Changing this library and recompiling the software will automatically generate a new version. (N.B. The object dimensions do not have to be a power of two). Some other possible sizes which will give about the same performance are : 128*128*128, 256*256*32 or 64*128*256.

- A trade-off between system performance and cost is very easy, because of the modular set-up. The software is identical for all Subcube processors, parameters are used to indicate which actual slices are to be stored and processed on a specific node.

15.7 Current activities

Work on the Voxel-processor prototype is continued in a number of areas :

- Addition of more 3D image-processing algorithms. An important feature will be the computer assisted image segmentation (region growing) to select interesting areas in the voxel-image.

- Implementation of 3D geometrical measurements. For medical- and biological- imaging geometrical data is very important. Surface computations, distances and volume measurements have to be applied to the objects in the voxel-space.

- Increase system performance by further code improvement and architecture optimization. For larger voxel-data sizes a system with 32 or more Transputers could be used. In this larger system the architecture will be changed to a tree structure. The advantage of a tree over a pipeline is the shorter average length of the communication path between the PE's and the controller.
- Investigate (voxel) data-compression, determine effects on data transport times and implications on transform algorithms.
- Feasibility study on stereoscopic display facilities.

15.8 Conclusions

The Voxel-processor is a successful demonstration of the performance improvement and flexibility that parallel processing can deliver. Transputers have proved to be a very powerful tool, both for research and applications. The development of the system software was greatly simplified by the clear representation and support of parallelism that OCCAM offers.

The key features of the developed Voxel-processor are :

- Fast, interactive system. Typical rendering speeds are 1 sec. for 2 Mbyte voxel-images. The speed may be increased by using more processors.
- Highly modular and easily adaptable software. The prototype is a general purpose framework for 3D image processing.
- Low-cost, small-sized off-the-shelf hardware. Transputers are general purpose processors and the system may also be used for other (computational intensive) applications.
- Flexible performance (linear cost/performance function).

15.9 References

[1] S. M. Goldwasser and R. A. Reynolds. Real-Time Display and Manipulation of 3D Medical Objects : The Voxel Processor Architecture. *Computer Vision, Graphics and Image Processing*, 39:1–27, 1987.

[2] C.A.R. Hoare. *OCCAM 2 Reference Manual.* Prentice Hall, 1988.

[3] INMOS. The Transputer Databook. Technical report, INMOS, 1989.

[4] A. Kaufman and R. Bakalash. Memory and Processing Architecture for 3-D Voxel-Based Imagery. *IEEE Computer Graphics and Applications*, November 1988.

16 Spatial Editing for Interactive Inspection of Voxel Models

G. J. Jense and D. P. Huijsmans

ABSTRACT
Voxel models are suitable for the representation of three dimensional objects of arbitrary topological complexity. They are mostly used for storing spatially sampled real-world data or data resulting from scientific simulation programs. In order to bring out the possibly highly irregular structure of the volume data, a visualization system for voxel-based objects should not only offer various (surface- or volume-) rendering methods, but also spatial editing operations.
We propose using an editing method, based on binary space partitioning. Construction of the binary space partitioning tree, that represents the subdivision of the voxel model, is done by interactive steering of the partitioning planes through the voxel model. The resulting BSP-tree is subsequently used in the rendering of the object. The advantage of a BSP-tree based partitioning is that it may be used in conjunction with many existing volume and surface rendering algorithms.

16.1 Introduction

Volume data may be produced by various types of acquisition equipment:

- Computer assisted tomography (CAT), magnetic resonance (MR), and ultrasound (US) scanners provide three dimensional data for medical diagnostic purposes, organized as series of parallel virtual slices, where each slice typically consists of 256×256 or 512×512 8- or 12-bit greyvalue elements.
- Series of microscope photographs of physical slices or *coupes* from a biological specimen. Usually, regions of interest in each coupe are marked by sets of, possibly nested, closed contours.
- Confocal scanning laser microscopes (CSLMs), where the extremely small depth of field of these devices is used to make virtual slices through a sample of tissue. The resulting dataset may consist of up to 512 3 8-bit samples.
- Analysis programs for seismic data yield 3D models of geological structures from which geologists try to deduct the location of, e.g., oil-, gas-, and waterreservoirs.
- Simulation programs for various physical phenomena, such as those used in atmospheric research, usually provide numeric data on some 3D grid.

All the above mentioned methods produce a so-called *voxel model* of the sampled physical object, i.e., an exhaustive enumeration of the occupancy of elementary volume cells on a regular 3D grid. Analoguous to traditional 2D image processing, voxel models may be perceived as three dimensional digital images. The fundamental difference with 2D image processing is that an *explicit* display operation is required before the contents of the 3D image can be visualized.

The remainder of this paper is structured as follows: we first give an overview of existing methods for the visualization of volume data. We then review the origins of the BSP-tree in section 16.2. Section 16.3 shows how existing volume rendering algorithms can be used in a BSP-tree based approach for the display of subdivided volume data. Subsequently, we describe our method for the interactive construction of a BSP-tree in section 16.4. Several details of the implementation of a volume visualization tool, based on the described techniques, are given in section 16.5, and some results are shown. Finally, we draw some conclusions and give directions for further work.

16.1.1 Volume visualization

In order to gain insight in the possibly highly irregular spatial structure that may be contained in 3D images, the volume data must be adequately visualized. Not only is it desirable to be able to display an irregular object, but also to manipulate it interactively. This allows one to visualize a topologically complex object in an explorative way. For that purpose it is necessary that one obtains rapid screen feedback of the initiated actions.

The size of volume data sets is such that conventional display methods are often not capable of generating an image within an acceptable time. Most display algorithms are based on *surface descriptions* of 3D objects such as surface tessalations with triangular (or generally: polygonal) patches. Thus, the required surface description must first be extracted from the volume data set in a separate processing step [1, 15, 9].

Traditionally, researchers in medical and biological fields, have attempted to visualize complex 3D objects by rapidly displaying the volume data slice by slice, from back to front and vice-versa, and mentally connecting the features of interest. Reslicing a volume data set along an arbitrary direction, different from the original slicing planes, may provide additional information about the internal structure of the object [11].

16.1.2 Volume rendering methods

Generating a 2D image directly from the volume data seems to be a better way. At present, several research groups all over the world are developing methods for volume visualization [4]. This has resulted in several *volume rendering* methods, i.e., algorithms that display particular features of interest directly from volume data sets. These algorithms may be categorized in three broad classes:

1. Forward mapping methods. Algorithms in this class project voxels from object space onto screen pixels in image space. This involves calculation of a viewing transformation matrix and applying that to the coordinates of all voxels that are to be projected. A well known algorithm that falls in this class is the Back-To-Front algorithm[5]. The Back-To-Front algorithm traverses the voxel array in a sequence, determined by the location of the viewpoint, that guarantees that voxels closer to the screen are projected later than voxels that are further away, thereby overwriting previously projected voxels.

When the above mentioned algorithm projects voxels from object space to image space, it performs a *binary classification* of the voxels in the data set, i.e., a yes-no decision on whether to project a voxel on the screen. This may lead to artifacts in the rendered image. Such a classification may be done by means of grey value thresholding. Binary voxel classification effectively yields a *surface rendering* of the opaque volume, consisting of the voxels with values above the selected threshold.

Other forward mapping algorithms provide true volume rendering by taking transparency into account in some way. The display algorithm of the Pixar volume rendering system [3] employs a sophisticated voxel classification scheme. This method considers voxels as containing certain percentages of different materials. From these percentages, colour and opacity values are determined, as well as density gradients. The colour and opacity values are used to render volumes of translucent materials, while the gradient vectors are used to render (transparent) surfaces that mark transitions between materials.

Although the above mentioned algorithms differ in many aspects, they have one important thing in common: the object space data (voxels) are traversed in a regular, usually slice-by-slice, row-by-row, column-by-column ordered, fashion. This makes these algorithms good candidates for either implementation in hardware, or for support from pipelined or parallel architectures.

2. Backward mapping methods. Instead of projecting voxels from object space to image space, backward mapping algorithms operate the other way around: for each screen pixel, they determine which voxels project onto them. This is usually accomplished by some type of *raycast* algorithm: rays are fired from the viewpoint through the screen pixels towards the volume model. A search is made along each ray and the values of the voxels encountered determine the pixel's colour. Backward mapping algorithms can be used to render both opaque surfaces and transparent volumes.

The simplest use of raycasting for the display of volume data is detection of the visible surface of a binary voxel model. This merely involves searching along each ray for the first 1-voxel [19]. A more sophisticated surface detection algorithm, that also takes a small region around the voxels on the ray into account, is described in [18].

Visible surface detection in greyvalue voxel models may be performed by, e.g., simple greylevel thresholding, or determining the greylevel gradient, based on the voxel values in a local neighborhood around the voxels on the ray [10].

A backward mapping strategy that avoids binary classification of voxel values, is described in [14]. The algorithm first collects voxel colours and opacity values along a ray and then composes a pixel colour from these values.

True volume visualization algorithms must in one way or another handle transparency. A method that integrates voxel values along each ray and stops when the summed value reaches a predetermined maximum, is called *additive reprojection* [10]. Thin features, with signifigantly larger voxel values than the surrounding voxels, are lost, due to the averaging along the ray that is performed in this method. Such features may be rendered more clearly by detecting the *maximum voxel value* along a ray [12].

3. Slicing. The ability to move a plane through a volume data set and rendering the voxels that are intersected by it, provides a way to visualize the internal structure of a voxel model. This method is especially effective when the plane can be interactively "steered" through the volume [12]. This allows the user to associate the changing image with the movements of the interaction device and thus mentally reconstruct the three dimensional structure of the object. By using multiple planes the user can delimit a (convex) subpart of the volume data. This method is known as *multiplanar reprojection*, or *multiplanar reformatting*.

When generating an image of the voxels that are intersected by a slicing plane through a volume data set containing N^3 voxels, only in the order of N^2 voxels have to be accessed, as opposed to about all N^3 voxels for volume rendering algorithms. This makes slicing algorithms potentially fast.

16.1.3 Spatial editing

It is likely that the nature of the underlying physical phenomenon, from which the volume data set originated, determines to a large extent the choice of a particular visualization technique. For instance, a more or less amorphous distribution of some scalar variable may be displayed, using an method that employs transparency. However, when there are clearly delimited substructures present in the data, it is often desirable to extract these substructures and manipulate them separately. As automatic segmentation of voxel models (which may, after all, be described as 3D digital images) can be difficult, an obvious solution is to create a subdivision of the voxel model through interactive manipulation. Such a spatial editing facility would allow a user to remove parts of the volume data in order to reveal hidden details. Systems that allow spatial editing in some form have been described before [2, 4], but their methods are rather limited.

One way to create a quite general subdivision of a voxel model is by means of recursively bisecting the voxel model with partitioning planes. The user may place these planes at arbitrary positions and at any orientation through the voxel model. The resulting subdivision of the voxel model (in convex parts) is described with an auxiliary data structure: the *binary space partitioning tree*, or BSP-tree. The voxel model can subsequently be visualized by displaying these parts separately, translated with repsect to a common centre (exploded view). A BSP-tree based editing scheme allows the incorporation of forward mapping, backward mapping, and slicing methods to render the volume data.

16.2 BSP-tree fundamentals

Originally, the binary space partitioning tree was devised to speed up the display of 3D objects, represented by a collection of polygonal boundary surface elements [7, 6]. BSP-trees have also been used in the evaluation of set operations on polyhedral objects [17], and in ray-tracing algorithms [16]. These applications all require some sort of spatial pre-sortedness of the polygonal input data set.

The spatial sorting of the input data is achieved by constructing a tree:

1. Choose a polygon from the input data set and place it at the root of the tree. Determine (the equation of) the infinite plane that embeds or supports the polygon.
2. Partition the remaining polygons in two subsets:
   (a) those that lie entirely to the left of the partitioning plane,
   (b) and those that lie entirely to the right of the partitioning plane,
   (c) Polygons that are intersected by the partitioning plane, are split along the intersection and each part is added to the appropriate subset.
3. Recursively apply the previous steps to the two subsets, until the input data set is exhausted.

The choice of which side is "left" and which is "right" is arbitrary. It is usually chosen to correspond with the counterclockwise and clockwise orientation of the polygon vertex list. Figures 16.1 through 16.4 give a simple example of a cube, representing a part of 3D space, subdivided by several planes, along with the associated BSP-tree.

Seen in another way, the planes, associated with the polygons, partition 3D space in several, possibly half open, convex subvolumes or *cells*. Thus, internal nodes of the BSP-tree represent partitioning planes, $ax + by + cz + d = 0$, which splits 3D space in two halfspaces, represented by its two children:

1. $\{(x, y, z) \mid ax + by + cz + d > 0\}$,
2. $\{(x, y, z) \mid ax + by + cz + d < 0\}$,

while leaf nodes represent convex volume cells, formed by the intersection of all halfspaces encountered on the path from the root of the tree to that leaf.

Note that the BSP-tree is a static data structure: it is constructed once for a given input data set and individual polygons or partitioning planes can not be dynamically

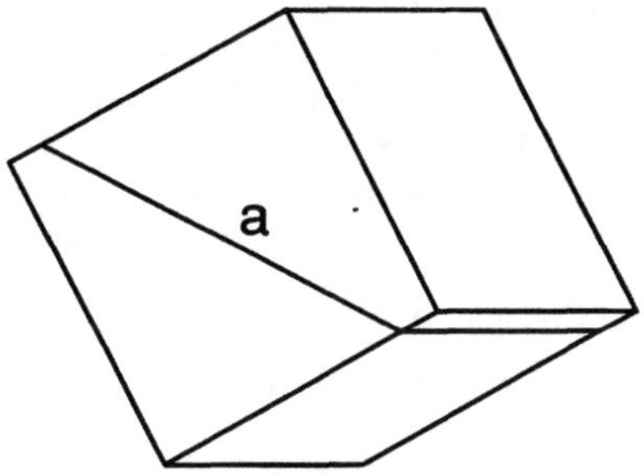

FIGURE 16.1. Cube, partitioned by one plane.

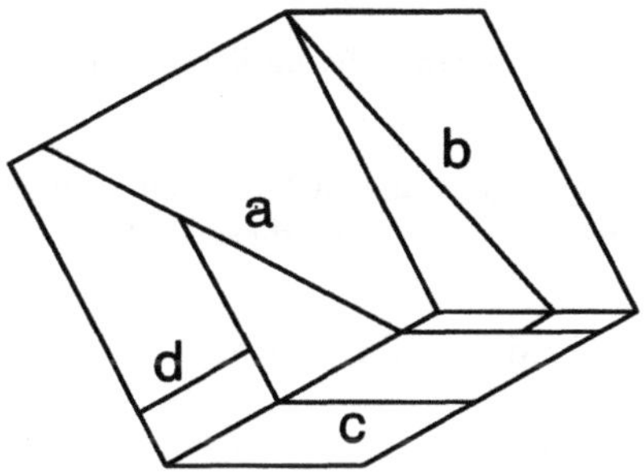

FIGURE 16.2. Several more partitioning planes.

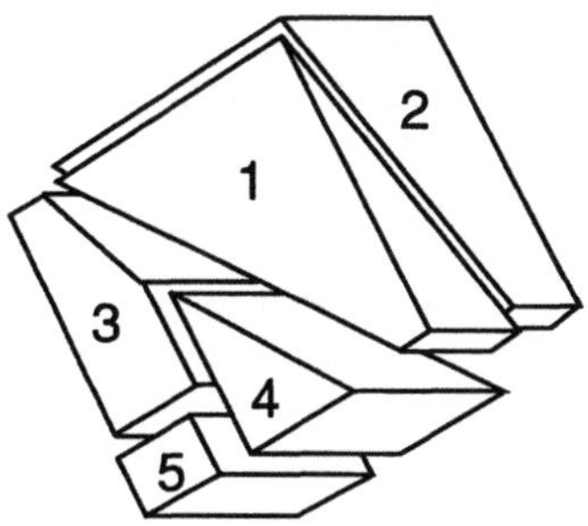

FIGURE 16.3. Cells shown in exploded view.

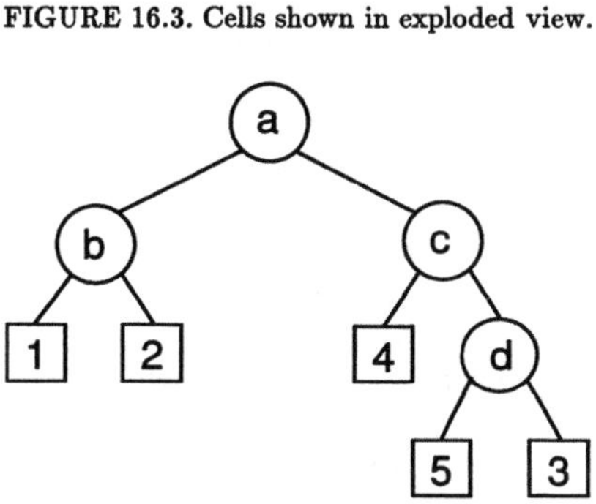

FIGURE 16.4. Corresponding BSP-tree.

inserted or deleted. In case the input data set changes, e.g. when a new partitioning plane or polygon is added or one is removed, the whole tree must be rebuilt from the internal node, where the new polygon is inserted, up. The deletion of internal nodes may even lead to a complete reordering of the tree.

Displaying an image from the constructed BSP-tree is straightforward: given the position of the viewpoint, the BSP-tree is traversed in the following order:

1. At a node, determine the side of the polygon the viewpoint is on.
2. Process the subtrees of this node in the order:
   (a) Traverse the subtree on the other side of the polygon.
   (b) Display the polygon at the current node.
   (c) Traverse the subtree on the same side of the polygon.
3. Apply the previous steps recursively, until the root is reached from its subtree at the "same" side.

This algorithm displays the polygons in a back-to-front sequence, always overwriting polygons on the screen that are further away by those that are closer to the viewpoint.

16.3 Displaying subdivided volume data

In the previous section we showed how a subdivision of (a part of) 3D space may be described by means of a BSP-tree. We will now describe how the standard BSP-tree display algorithm can be adapted to obtain an *exploded view* of a subdivided voxel model.

The BSP-tree display algorithm involves an in-order tree traversal yielding the polygons of the input data set in a back-to-front sequence. When displaying a subdivided voxel model, the BSP-tree describing the subdivision is traversed in the same way, but at each node, different display actions are taken, depending on the type of the node (*leaf* or *internal*) and the values of its attributes.

The advantage of using a BSP-tree here, is that it represents both the partitioning of the volume data in convex cells, as well as the polygons that separate these cells. Forward mapping volume rendering algorithms can employ the back-to-front ordering of the partitioning planes, as well as the convexity of the volume cells, while backward mapping algorithms benefit from the front-to-back ordering of the partitioning planes to speed up ray-polygon intersection tests.

A slicing algorithm is especially easy to add to the standard BSP-tree display algorithm. As a first step towards a more comprehensive editing system we have used this technique to implement an extended form of multiplanar reformatting to visualize the volume data.

In a standard BSP-tree, the internal nodes would hold a representation of a partitioning plane, for instance the coefficients of the plane equations, while the leaf nodes are empty, indicating that there is no further subdivision of space below this level. The polygons that make up the boundary surface of a convex volume cell at a leaf node, would then have to be determined at display time. This involves traversing the BSP-tree from the root to the designated leaf node and computing the intersection of all halfspaces encountered on this path, finally yielding a polygonal boundary surface when the leaf node is reached.

For our "augmented" BSP-tree, we calculate the polygonal boundary surfaces of the convex subvolumes once, during construction of the BSP-tree, and store them as a linear list of polygons at the corresponding leaf node. This approach is advantaguous when the subdivision of the volume model changes less frequently than the viewpoint from which an image is displayed. Another possibility would be not to store these boundary polygons explicitly, but to compute them "on the fly" during traversal of the tree, but this would increase display time. Using the syntax of the C language, a BSP-tree can be described with the following elements:

```
typedef struct vertex  Vertex;
typedef struct polygon Polygon;
typedef struct bspnode Bspnode;

struct vertex
{
  float   x, y, z;
};

struct polygon
{
  float   a, b, c, d;
  int     nverts;
  Vertex  *vlist;
  Polygon *next;
};

struct bspnode
{
  float   a, b, c, d;
  int     npolys;
  Polygon *plist;
  Vertex  centroid;
  int     vsblty;
  Bspnode *left, *right;
};
```

When the node is an internal one, the coefficients $a, b, c,$ and d of the plane equation are stored, and the pointers to the node's children are non-NULL. The other fields, in this case, are of no significance. Conversely, when the node is a leaf, the fields *npolys* and *plist* give the number of polygons and the first element in the polygon list respectively. Each leaf node also has a *visibility* attribute, indicating whether the polygons of that cell are to be displayed or not. The *centroid* field contains the value of the arithmetic mean of the polyhedron's vertices. Assuming the polyhedron consists of N vertices $v_1, \ldots, v_N$, the centroid is $(v_1 + \ldots + v_N)/N$. Its use will be clarified later.

The display algorithm must also be changed to yield the leaf nodes, instead of the internal nodes, in a back-to-front order:

1. Starting at the root node, determine the nodetype (*internal* or *leaf*). If the node is a leaf, render the visible polygons from its polygon list, else
2. Visit the nodes of the subtree, rooted at this internal node, in the order:

   (a) Traverse the subtree *behind* the partitioning plane.

   (b) Traverse the subtree *in front of* the partitioning plane.

Our version of the BSP-tree display algorithm is a pre-order tree traversal (instead of an in-order traversal) because in the original BSP-tree the polygons are stored in the

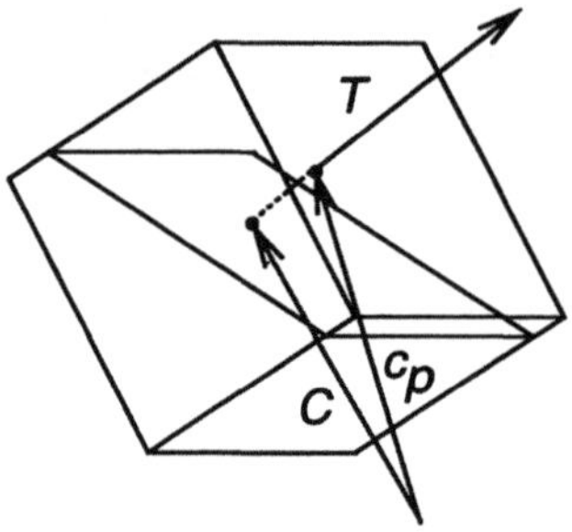

FIGURE 16.5. Determining the translation vectors.

internal nodes, while we store them in the leaf nodes. Also note that some information in our BSP-tree is redundant: the coefficients of the plane equations in the internal nodes occur again in the polygon lists of the leaf nodes. The availability of these data in two places simplifies various calculations at the cost of a small penalty in memory use.

As stated before, our aim is to bring out hidden details in a voxel model by allowing a user to remove certain parts of the volume data. This can be done by setting the visibility attribute of various leaf nodes in the BSP-tree to OFF. Another way to reveal the contents of the voxel model, while at the same time retaining the spatial relationships between different substructures is to provide an *exploded view* facility.

An exploded view is obtained by translating all volume cells along a vector $\vec{T_e}$ by a certain amount, outward from the "main" center of the voxel model $\vec{C}$. The amount of translation is determined by the distance of the polyhedron's centroid, $\vec{c_p}$, to the center of the voxel model, multiplied by a constant factor f that may be set by the user. The direction of translation is from the main center to the cell's centroid:

$$\vec{T_e} = f(\vec{c_p} - \vec{C})$$

Thus, polyhedra that are further outward from the main center are translated by a larger amount than those closer to the main center. This, together with the fact that all cells are convex, guarantees that there can be no "collisions" between translated cells (see also figure 16.5).

We have now shown how we use a BSP-tree based subdivision in two ways to spatially edit voxel models:

1. by allowing cells to be marked *invisible*, details of an object that are behind it can be revealed,
2. by translating all cells outward (exploded view), hidden parts become visible, while the spatial relationships between cells are preserved.

Each method, in its own way, offers a different way to visualize the internal structure of the volume data. The combination of these two methods is of course also possible.

16.4 Interactive BSP-tree construction

Construction of a BSP-tree proceeds as follows: the root of the BSP-tree represents the part of 3D space, occupied by the voxel model. This root cell is split in two by a partitioning plane, resulting in two sibling cells, *front* and *back*. The user may select either of

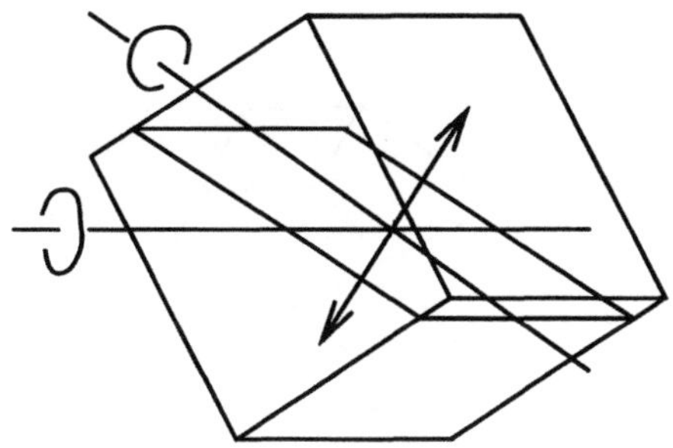

FIGURE 16.6. Directions of movement for a partitioning plane.

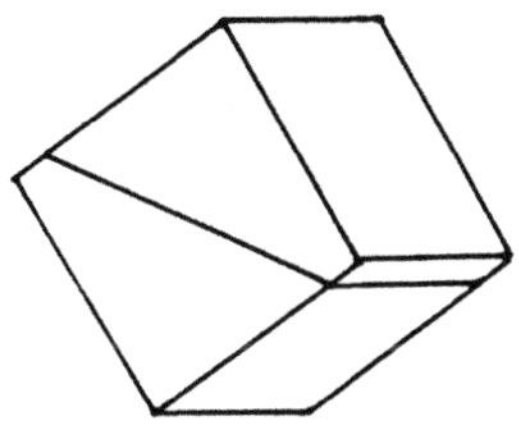

FIGURE 16.7. Initial subdivision.

these for further subdivision, making it the *current* cell. These actions may be repeated until the desired subdivision is reached.

Positioning a plane is done by manipulating three sliders on the screen by means of a mouse. Two of these sliders determine rotation angles about X and Y axes in the plane (these angles are sometimes called *yaw* and *pitch*, respectively). Dragging a third slider translates the plane along its normal vector (figure 16.6).

The system offers two display "modes" during the editing operation:

1. showing only the currently selected cell,
2. showing the current cell, its parent in the BSP tree, and its two children.

This latter mode provides the user with information about the spatial relationships between cells in the neighborhood of the current cell.

Splitting the part of 3D space, represented by the current cell, can be also be done in two ways:

1. a new partitioning plane splits *all* cells that were created in previous subdivisions,
2. a new partitioning plane splits only the *current* cell in two new subcells.

Figures 16.7 through 16.8 illustrate both partitioning methods. In either way, more and more partioning planes may be added, until the desired subdivision of the voxel model is reached. Each of these two methods has its own advantages. Which method is used, mostly depends on the contents of the volume data set.

In order to relate the positions of the partitioning planes to the volume data, some form of visual feedback must be presented to the user. Real-time volume rendering of the voxel model, using either one of the described forward or backward mapping algorithms,

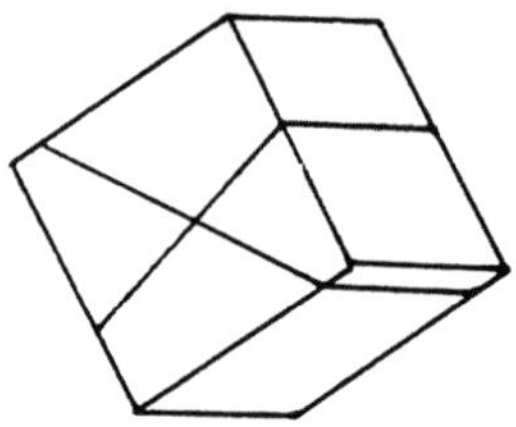

FIGURE 16.8. All cells are split.

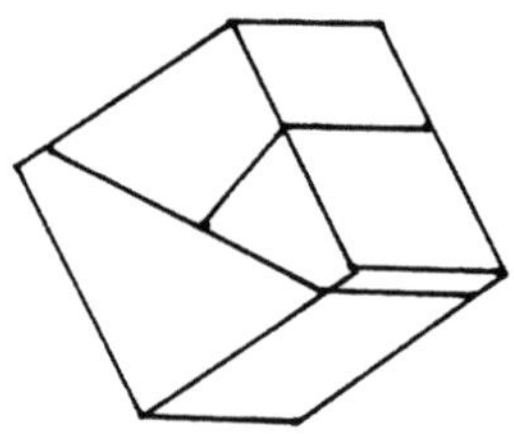

FIGURE 16.9. Only the current cell is split.

is currently not feasible. We therefore employ a slicing algorithm as a sensible alternative. The outline of the voxel model is displayed as a wireframe cube (or parallelepiped) and an initial partitioning plane is displayed at a starting position (see colour plate 49).

Now, the user may alter the position of this plane by translating and/or rotating it, until the desired position and orientation are reached. The position of the plane w.r.t. the volume data is shown by voxel-mapping the corresponding polygon, i.e. for each screen pixel of the polygon, determining which voxel projects onto it and rendering its value as a pixel colour. This slicing method is fast enough to enable interactive positioning of the plane through the voxel model.

As stated in section 16.2, a BSP-tree is a static data structure. This means that when a new partitioning plane is added, a new BSP-tree has to be constructed. Because N partitioning planes may potentially yield a subdivision of 3D space, consisting of N^3 convex cells, one would expect this to be a costly operation in terms of processing time. However, in practice a BSP-tree, constructed from N partitioning planes, will contain CN cells, with C a constant between 1 and 5 [6].

16.5 Implementation and results

The available hardware platform for the implementation of our "voxel editor" consists of a Sun 3/160 workstation, equipped with a TAAC1 accelerator board. The user interface and the routines related to the BSP-tree, as well as the file I/O parts of the programs, all run on the Sun host computer. The TAAC1 is used to store the volume data and execute the display routines.

Preliminary implementations have been made of a "BSP-tree editor" and of a program

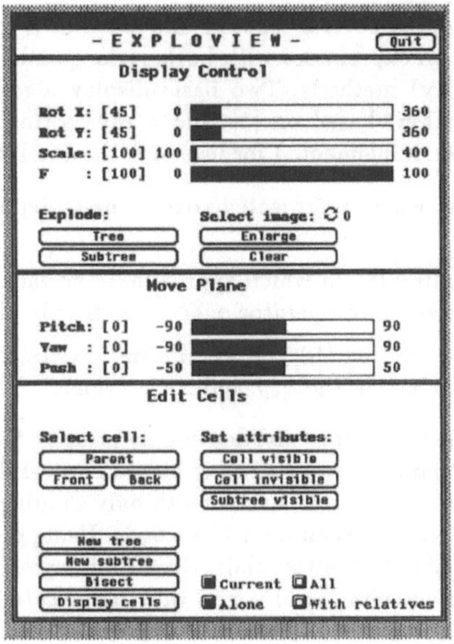

FIGURE 16.10. *ExploView* user interface

for experimenting with volume rendering algorithms. For both programs, a volume data set is available that was acquired with a CAT scanner. The voxel model consists of 128^3 8-bit voxels. This model was reconstructed from 128 CAT scans, each consisting of 256^2 12-bit pixels.

The BSP-tree based volume editor, *ExploView*, offers facilities to construct a BSP-tree based subdivision of a voxel model. The user interface groups screen devices into four categories:

control: selection of display and edit modes,

view: for setting the viewing parameters, such as the position of the viewpoint, scale factor and explosion factor f,

planes: sliders for positioning the current partitioning plane etc.,

cells: for moving among cells in the tree and setting node attributes.

The most costly operation (in terms of computational power) is the display of a plane as it is moved through the voxel model. This involves computation of the polygon that forms the intersection of the plane with the polyhedral surface of the current cell, and subsequently rendering this polygon with a slicing algorithm. The intersection computation is performed by the host computer, while the rendering operation is performed by the accelerator board. This leads to a display speed of about 5 images per second, which is fast enough for interactive construction of the subdivision.

For the generation of an exploded view, the tree-traversal algorithm on the host machine yields the visible polygons in a back-to-front sequence. These are then passed in this order to the accellerator board for rendering. An exploded view of a voxel model that has been subdivided into approximately 40 cells is generated in less then 1 second. This includes the rendering (using the slicing method) of between 100 and 150 visible polygons.

We have also implemented several forward and backward mapping volume rendering methods. This was done to experiment with both surface and true volume rendering (i.e. using partial transparency) methods. Two basic display algorithms were implemented: the Back-to-Front algorithm, based on [5], and a ray-casting algorithm. Three surface rendering techniques were implemented for use with the Back-to-Front algorithm:

1. depth-only shading, where a projected voxel's shade depends just on its distance to the viewing plane,
2. depth-gradient shading [8], in which approximate surface normal vectors are calculated by running a gradient operator over a depth-only shaded pre-image,
3. grey-value gradient shading [10], i.e. computing normal vectors from the local gradient of the voxel values in the $x, y,$ and z directions.

The raycast algorithm currently does just depth-only shading (possibly followed by depth-gradient shading, in a separate post-processing step). Typical display times for the Back-to-Front algorithm are: 15–30 seconds, using depth-only shading, while the depth-gradient shading post-processing step takes another 10 seconds. Using grey-value gradient shading, an image is generated in about 20–40 seconds. Display times vary with parameter settings, and generally depend on the number of voxels that are selected from the voxel model for projection and rendering. Figure 16.11 shows images that result from different rendering rendering methods. These images also illustrate the differences in the amount of detail shown. The size of all images is 512×512 pixels.

16.6 Conclusions and further work

Our experience with the BSP-tree editor indicates that interactive construction of a subdivision of polyhedral volume cells is feasible. The timing results from the display algorithms show that rendering an image from a full 128 3 voxel data set takes too long for a truly interactive system. However, display times can be reduced by creating a subdivision and marking cells, containing irrelevant data, as *invisible*. Thus, they no longer contribute to the rendering time.

A feature that has proved to be very useful is the ability to render an image at reduced resolution. In the Back-to-Front algorithm, this is accomplished by stepping through the voxel model with strides 2 or 4, thus displaying a 64 3 or 32 3 voxel model instead of the full model. The display times at these resolutions are 5 seconds and less than 1 second respectively, which makes the setting of global viewing parameters less tedious.

Integration of the display algorithms in the BSP-tree editor is now under way. This involves adaptations of both forward and backward mapping methods to take advantage of the spatial organization of the volume cells in the BSP-tree.

- The original Back-to-Front algorithm was designed for the traversal of voxels, contained within a rectangular box. The algorithm will therefore have to be extended to handle convex polyhedral boundaries. An algorithm for the "3D scan-conversion" of such objects is given in [13]. This algorithm will be adapted for use in our system.
- The raycasting algorithm will be extended so that it benefits from the spatial sortedness of the polygonal surfaces for determining ray-cell intersections [16].

In addition to the visibility attribute, each leaf node will also contain an attribute that specifies the rendering method that is to be applied to that volume cell. This attribute may have values such as *depth-only*, *greyvalue-gradient*, and *transparent*. Additionally each

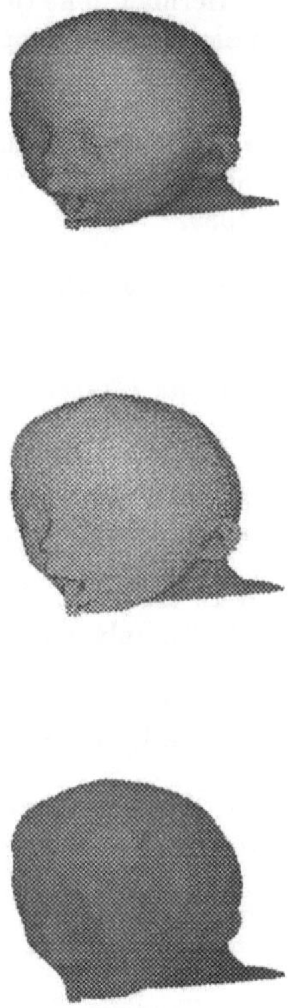

FIGURE 16.11. Depth-only, depth-gradient and greyvalue-gradient renderings of a volume data set

internal node will also have a visibility attribute, which controls whether the corresponding polygon is to be voxel-mapped at display time, or not.

A final improvement will be the development of a better user interface. The steering of the partitioning planes would especially benefit of some form of direct manipulation facility instead of indirect manipulation through sliders.

In conclusion, the BSP-tree based subdivision scheme provides sufficiently "rich" spatial editing facilities to serve as the basis for a system for visualizing voxel models. It allows the application of a wide range of volume rendering methods and combines them to provide truly interactive inspection of volume data.

16.7 References

[1] E. Artzy, G. Frieder, and G. T. Herman. The theory, design and evaluation of a three-dimensional surface detection algorithm. *Computer Graphics and Image Processing*, 15(1), 1981.

[2] L.-S. Chen and M. R. Sontag. Representation, display, and manipulation of 3D digital scenes and their medical applications. *Computer Vision, Graphics, and Image Processing*, 48(2), November 1989.

[3] R. A. Drebin, L. Carpenter, and P. Hanrahan. Volume rendering. *Computer Graphics*, 22(4):65–74, August 1988.

[4] C. Upson (ed.). *Proceedings of the Chapel Hill workshop on volume visualization.* ACM, May 1989.

[5] G. Frieder, D. Gordon, and R. A. Reynolds. Back-to-front display of voxel based objects. *IEEE Computer Graphics and Applications*, 5(1):52–60, January 1985.

[6] H. Fuchs, G. D. Abram, and E. D.Grant. Near real-time shaded display of rigid objects. *Computer Graphics*, 17(3), July 1983.

[7] H. Fuchs, Z. Kedem, and B. Naylor. On visible surface generation by a priori tree structures. *Computer Graphics*, 14(3), June 1980.

[8] D. Gordon and R. A. Reynolds. Image space shading of 3-dimensional objects. *Computer Vision, Graphics and Image Processing*, 29(3), 1985.

[9] D. Gordon and J. K. Udupa. Fast surface tracking in three-dimensional binary images. *Computer Vision, Graphics and Image Processing*, 45(2), February 1988.

[10] K. H. Höhne and R. Bernstein. Shading 3D images from CT using gray level gradients. *IEEE Transactions on Medical Imaging*, 5:45–47, March 1986.

[11] D. P. Huijsmans, W. H. Lamers, J. A. Los, and J. Stracke. Toward computerized morphometric facilities. *The Anatomical Record*, 216:449–470, 1986.

[12] E. R. Johnson and C. E. Mosher. Integration of volume rendering and geometric graphics. *Proceedings of the Chapel Hill workshop on volume visualization*, May 1989.

[13] A. Kaufman and E. Shimony. 3D scan-conversion algorithms for voxel based graphics. *Proceedings of the ACM workshop Interactive 3D graphics*, October 1986.

[14] M. Levoy. Display of surfaces from volume data. *IEEE Computer Graphics and Applications*, 8(2):29–37, May 1988.

[15] W. Lorensen and H. Cline. Marching cubes: A high resolution 3D surface construction algorithm. *Computer Graphics*, 21(4):163–169, July 1987.

[16] B. F. Naylor and W. C. Thibault. Application of BSP trees to ray-tracing and CSG evaluation. Technical Report GIT-ICS 86/03, School of Information and Computer Science, Georgia Institute of Technology, Atlanta, Georgia 30332, USA, February 1986.

[17] W. C. Thibault and B. F. Naylor. Set operations on polyhedra using binary space partitioning trees. *Computer Graphics*, 21(4), July 1987.

[18] Y. Trousset and F. Schmitt. Active-ray tracing for 3D medical imaging. In *Eurographics 87*, pages 139–150, August 1987.

[19] H. K. Tuy and L. T. Tuy. Direct 2-D display of 3-D objects. *IEEE Computer Graphics and Applications*, 4(10):29–33, October 1984.

Part V

Interaction

17 The Rotating Cube: Interactive Specification of Viewing for Volume Visualization

Martin Frühauf, Kennet Karlsson

ABSTRACT
A part of the user interface of a volume visualization system is described. It provides the opportunity of the real-time interactive definition of viewing parameters for volume rendering. Viewing parameters in this case are the view point and cut planes through the volume data set. It uses an approach for the fast rendering of volume data which traditional computer graphics does not know and which is as fast as wire frame representations.

17.1 Introduction

The volume rendering of huge volume data sets in scientific visualization is very computing-intensive. High quality images from those data sets cannot be computed in or near real-time on general purpose graphic-workstations, not even on super-workstations. A special tool for the interactive specification of viewing parameters for volume rendering is thus required. Viewing parameters in this case are the viewpoint and the location of cut planes through the data set. The tool must provide the opportunity of orientation to the scientists even in huge data sets. The echo of every user interaction must be computed in real-time. In the following the term "Volume Rendering" is used for the rendering of volume data directly from volume primitives as applied e.g., in medical imaging. In animation systems, for instance, wire-frame representations are used to define the motion of objects interactively, while the final frames are rendered using these motion parameters afterwards. Wire-frame representations are also used in CAD systems during the construction of objects, whereas shaded representations are computed afterwards. In volume rendering the use of a wire-frame representation is not possible due to different reasons. The first reason is the lack of any explicit surface representation of objects in the data set. The second reason is that the interior of "objects" in the data set is not homogeneous. The neglect of that inhomogeneity as in wire-frame representation would complicate the orientation in the data set for the scientist. The third reason is that the structure and thus the surface of "objects" created by the interpretation of the data set is very complex. Therefore a wire-frame representation is difficult to compute and would in most cases consist of many vectors. Furthermore, surface representations of volume data have to be recomputed after slicing the data set. Due to these reasons we have developed a special tool for the interactive definition of viewing parameters, and we are using this tool with different volume renderers (plate 55) [2]. Another reason for the development of the rotating cube is the fact, that a special user interface in volume visualization systems is required for scientists who are not familiar with the principles of rotation, projection and lighting and shading in computer graphics (e.g., medical staff) [5, 6].

17.2 Concepts

In the following we describe the concepts and the implementation of a tool for the interactive definition of the viewpoint of scientific volume data and cut planes through such data, i.e., the rotation and cutting of volume data in real time. Volume data is mostly arranged in a regular grid, i.e., a data cube. The orientation of the cube is perceived by the user from the location of its vertices and edges. Back and front, left and right, bottom and top can be distinguished by the interior structure of the cube's surfaces. Therefore we project 2D pixmaps from the volume data set on the six surfaces of the cube. 2D pixmaps on a cube are sufficient for orientation, since most of the scientists are now used to evaluate their data sets with the aid of 2D images, and 2D images are the source of many scientific volume data sets. The simplest version is to map the data from the outer layers to the cube's surfaces. In case that these layers do not contain any data, a threshold depending on the user's interpretation of the data set above the data noise is specified. Data above this threshold is then orthogonally projected to the cubes surfaces. These six projections are performed in preprocessing steps. Only one new surface is computed at a time after a cutting operation through the data set, because cut planes are perpendicular to the coordinate axes of the volume space.

17.3 Implementation

17.3.1 User interface

The tool has two input modes: a rotating and a cutting mode. It can be switched with, e.g., a toggle button (plates 53 to 54). The viewing direction is selected by rotating the volume on the screen. The rotation is done with the mouse. Each time a mouse button is pressed, the volume is rotated a certain angle, e.g., left button = 1deg, middle button = 5deg, right button = 20deg. The rotation axis and direction is determined by the position of the mouse (figure 17.1). The window of the tool is divided into fields, corresponding to an intuitive understanding of the rotation of the volume. The middle left part of the window corresponds to the rotation "to the left" etc. Thus, with some few natural mouse operations the viewing direction is selected, without having to care about the way the coordinate axes are oriented, positive or negative rotation direction, etc. One slice of the volume is cut away by picking one of the visible faces of the volume. The picked face is sliced off, i.e., the outer voxel layer on this side of the volume is cut off showing the next voxel layer. In this way it is possible to walk through the volume in real-time and define the cutting planes for the final high quality display (plate 55).

17.3.2 Calculation of the echo

In order to accelerate the rendering of the echo, only the six faces of the volume are rendered, not the inner parts. The faces are kept in the main memory as two-dimensional pictures in the form of pixmaps, which are normally the outer voxel layer of the corresponding faces of the volume. In case that this will result in empty pictures, the pixmaps contain the parallel projections of an inner object on the faces (plate 56). For the rotation of the volume a rotation matrix [1] is used. In order to accelerate the process, only the eight vertices of the volume are rotated. For each rotation, i.e., each time a mouse button is pressed, the matrix is updated and the new positions of the vertices are calculated by multiplication with the matrix. At any given viewing direction only one, two or at most three faces are visible. Thus at most three of the six faces must be drawn. This, together with the fact that the volume is convex, eliminates the hidden surface problem. The visible faces are determined from their normal vectors. The faces with a positive z-coordinate (figure 17.1) of the normal vector are visible. The volume is displayed with a parallel projection. This is done simply by skipping the z-coordinates of the vertices of the volume; no calculations are necessary. Each visible face appears as a parallelogram on the screen.

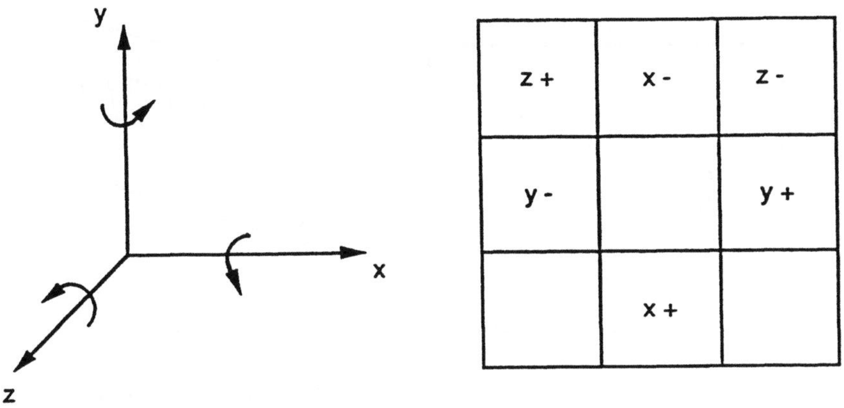

FIGURE 17.1. The window of the tool is divided into fields, corresponding to the rotation axes and directions

Through shearing and scaling of the pixmap the face is mapped onto the parallelogram (figure 17.2). The method used for the mapping is a scanline-based fill algorithm similar to the one presented in [4]. When a cut is performed, the selected face must be changed to the next deeper voxel slice in the volume or, if the faces represent projections of an object, the projection on the face being cut must be updated. If a Z-buffer of each face is kept, this updating of the projection can be done very fast. The adjacent faces of the face being cut must be narrowed by one pixel and the four vertices of the face must be moved by one voxel. In order to be able to update the face, - i.e., creating a new slice or a new projection - in real-time, the full volume must be held in the main memory. Therefore we reduce CT data sets of 2563 voxels by factor two on machines with less than 20 MB main memory. Our implementation enables a real-time manipulation of the volume. This is possible because of the following reasons:

- The rendering of the volume is done with simple two-dimensional pixmap-operations on at most three faces.
- While the rotation of a volume with e.g., 16 Mvoxels is a very CPU intensive work, we rotate only the eight vertices of the volume, reducing the effort to virtually nothing.

17.4 Conclusions

We have described our tool for the real-time interactive specification of viewing parameters for volume rendering. It is of great advantage in developing and applying our volume rendering techniques to various data sets, and we have found that it works very well. Nevertheless, this is just a part of the user interface of a volume visualization system. However the interactive rotation and slicing of huge data volumes in real-time on workstations is a great challenge and harder to solve than other parts of the user interface, e.g., colour assignment to volume data for rendering. On the other hand, a convenient tool for specifying the view point is essential for the scientists to explore their data. We

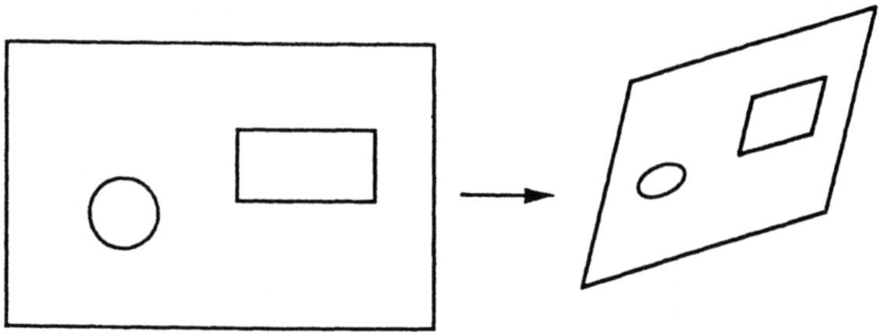

FIGURE 17.2. The face is mapped onto its parallelogram

will design and develop a complete user interface for volume rendering in the near future; the described tool will then be integrated in that user interface. In case of time-consuming rendering techniques, the described tool is already used on a workstation in a distributed system, whereas the rendering of volume data is performed on a supercomputer [3].

17.5 References

[1] J D Foley and A van Dam. *Fundamentals of Interactive Computer Graphics.* Addison-Wesley, 1983.

[2] M Frühauf. Volume Visualization on Workstations: Image Quality and Efficiency of Different Techniques. *Computer and Graphics*, 14(4), 1990.

[3] M Frühauf and K Karlsson. Visualisierung von Volumendaten in verteilten Systemen. In A Bode, R Dierstein, M Göbel, A Jaeschke, editors, *Visualisierung von Umweltdaten in Supercomputersystemen. Proc. GI-Fachtagung. Informatik-Fachberichte*, volume 230, pages 1–10. Springer, Berlin, 1989.

[4] G R Hofmann. Non-Planar Polygons and Photographic Components for Naturalism in Computer Graphics. In *Eurographics '89*, Amsterdam, 1989. North-Holland.

[5] H U Lemke et al. 3D Computer Graphics Workstation for Biomedical Information Modelling and Display. In *Proc. SPIE - Int. Soc. Opt. Eng.*, volume 767, pages 586–592, 1987. Conference Medical Imaging (SPIE), Newport Beach, CA, USA.

[6] Th Wendler. Cooperative Human-Machine Interfaces for Medical Image Workstations: A Scenario. In H U Lemke, editor, *Proc. Int. Symp.CAR'89*, pages 775–779. Springer, Berlin, 1989.

18 Chameleon: A Holistic Approach to Visualisation

N. Bowers, K. W. Brodlie

18.1 Introduction

18.1.1 Background

Scientists from beyond the field of computing are becoming increasingly aware of the advantages to be gained by visualising their problems. Not only does it increase productivity, but if used intelligently it can improve the user's understanding of the problem.

Although the end-user would prefer one visualisation system for all problems, attempting to provide for all perceived needs in one step would not only be unrealistic, but would probably result in a system of heterogeneous, rather than homogeneous, components. Therefore in designing a visualisation system we should aim for the following properties:

1. *Extensibility.* The ability to add functionality so that it integrates smoothly with the existing system.

2. *Flexibility.* The user must be able to modify the working environment to his or her taste, whether it be choice of background colour or the interaction style to be used.

3. *Usability.* The end users of the system should not need a degree in computer science to use it, but neither should they feel constrained by the interface. Very often a user interface becomes restrictive after the initial learning process, due to the designer interpreting 'easy to use' as 'simple'.

4. *Portability.* The user should not be constrained to one particular vendor or machine architecture. For many sites, portability is often an important factor.

In what has become a landmark report, McCormick *et al* defined the field of visualisation and outlined its objectives[6]. They noted that visualisation embraces a wide range of disciplines, which have previously been treated as independent research areas, including:

- Computer graphics
- User interface studies
- Scientific Computation

The Oxford Dictionary defines Holism as "*the tendency in nature to form wholes that are more than the sum of the parts by ordered grouping*". Our use of the word stems from the belief that scientists will gain a greater understanding of their problems if all aspects of the visualisation process, including the problem itself, are incorporated into one coherent system.

18.1.2 Area of interest

At Leeds we have been developing a system for visualising problems whose solution is a function defined over a region in space. The solution may be available directly as a function which can be evaluated at any point, or it may only be available at a discrete set of points. In the latter case, an interpolating function must be constructed from the data, and it is this function which then represents the solution.

Mathematically, the problem is to display

$$F(\mathbf{x}) \qquad \mathbf{x} \in \mathcal{R}$$

where $\mathbf{x} = (x_1, x_2, \ldots, x_N)$ and $\mathcal{R}$ is a region in N-dimensional space. The function F is assumed to yield a unique value at any point $\mathbf{x}$. Note that this is a subset of the more general multi-dimensional visualisation problem, where F is a vector-valued function. Nevertheless this present problem, with one dependent variable and many independent variables, is sufficiently broad to encompass many real-life problems (see next section).

Moreover, it is a challenging visualisation problem, particularly as the number of independent variables increases. It is generally impossible to show all aspects of the function in one display, or even a predefined sequence of displays. Instead we must allow the scientist to 'browse' or 'explore' the function interactively.

Interaction is seen as the key to effective visualisation. The user must not only have control over the visual aspects, but should also be able to direct the scientific computation. The user interface is therefore a critical component: if the scientist is to gain the understanding of the problem that we aim for, then the interface must be couched in domain specific terminology and imagery. Since no two users would agree on the definition of a 'best' user interface, the system should be chameleon-like in nature, with the interface adapting itself to the user, instead of the reverse, which is so often the case.

18.1.3 Example problem

Our work is aimed at the general problem just described, but it has been driven by a particular example from the field of chemistry where visualisation is needed. This example will be used throughout the paper to illustrate our ideas.

Very often it is important to know whether the vapour above a liquid mixture contains the components in the same ratio as the liquid. If this is true, the mixture is called an *Azeotrope*. A mixture of three components can be represented as a point within an equilateral triangle - the barycentric coordinates representing the proportions of each component. An *azeotropy function* is defined over the triangle, and a maximum of this function represents an azeotrope.

Computation of this function is a non-trivial task. It involves the simulation of the interaction between the liquid's components at the molecular level. A pure azeotrope can be found by a numerical optimisation procedure, but scientists are interested not only in the azeotrope itself, but in the behaviour of the function near an azeotrope. This behaviour will be best conveyed by some graphical representation of the function - hence the need for visualisation.

A more challenging visualisation problem arises when a fourth component is introduced into the mixture. Now the scientist needs to explore the behaviour of the azeotropy function within a regular tetrahedron, and the system must provide him with the interaction tools required. Typically the scientist will wish to view the function keeping one of the components fixed, thus reducing the dimension of the problem. But how does he select the subspace he wishes to view? And what happens if further components are introduced?

18.1.4 Structure of this paper

Our visualisation system is to be called 'Chameleon'. It is presently at the design stage, with a small prototype to test out our ideas.

The following section contains a quick overview of the ideas behind Chameleon. Section 18.3 contains details of the *method* concept and its implementation. The *view* concept is discussed in section 18.4, and the relationship between methods and views clarified. Section 18.8 contains definitions of the different types of configuration. In Section 18.9 the constituent parts will be pulled together and the Chameleon system discussed as a whole. Finally Section 18.10 gives conclusions and ideas for future work on Chameleon.

18.2 Overview

Until recently scientific computation and the display of its results were considered two separate processes, with the scientist often iterating over the two steps many times. Incorporating the two steps into one task will obviously increase productivity, but will also encourage experimentation.

Our aim is to provide an extensible visualisation environment where all components build on a common foundation and present the same user interface. Previous visualisation software has not fully exploited the facilities offered by workstations - as a simple example, many existing systems do not intelligently handle window resizing. Such aspects should not be the concern of the problem owner or application programmer.

Users of Chameleon have to provide one or more problems which they wish to investigate. Problems are defined either as sets of numerical data, or as real-valued functions. The user should then be able to explore the problem at will, looking at it (or parts of it) from different perspectives, and perhaps modify the problem itself. In the introduction we mentioned that we are trying to meet *perceived* needs. We cannot hope to determine all user requirements of such a system, and users' expectations are always changing, so we must work towards a modular and flexible design.

Chameleon contains a library of techniques, or *methods*, for presenting information. Methods are made available to the user through *views*, which provide the mechanism for interacting with the method. Users can simultaneously visualise the same problem in many different ways, or can visualise different problems concurrently. Referring to our example problem, the azeotropy function for a three component liquid mixture could be displayed using a filled area contour method. Figure 18.1 shows a view containing an instance of just such a method.

18.3 The method concept

Within Chameleon, a *method* is a technique used to present information to the user. Whilst this could mean a surface plot, or histogram, it could also, for example, be an invocation of an editor on a data file. The definition of a method is intentionally abstract in order that the design not be constrained to traditional methods for displaying data.

Chameleon caters for two types of method, which we have called *internal* and *external* methods. Internal methods are based entirely on Chameleon's mechanisms, whilst external methods use some external graphics package. Both types of method can have properties which may be parameters of the display algorithm, such as a region of interest. Each method has an associated *descriptor* which contains declarations of properties - their type, default values etc. Chameleon defines a protocol with which the methods and Chameleon can pass values for the properties.

18.3.1 Internal methods

Chameleon's internal methods follow an object oriented approach, which encourages a modular and extensible design, and allows similar methods to share common code and

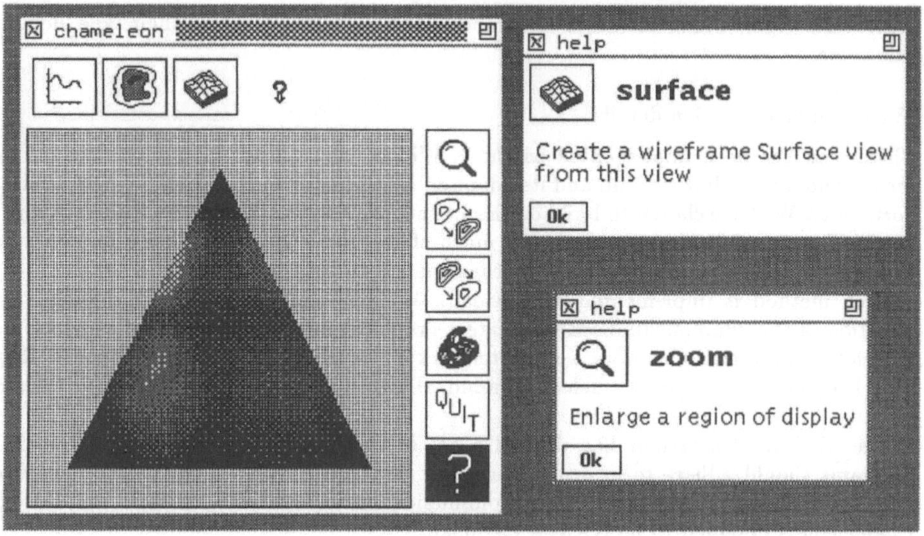

FIGURE 18.1. An example view, containing a contour plot method

data. For example, there will only be one routine for drawing axes, which is shared by all methods. This will also facilitate the provision of image capture mechanisms, for inclusion of pictures in reports etc.

Methods are organised in a class hierarchy, with new methods being subclassed from existing ones, inheriting the properties required in the new class. The base class in Chameleon is the **CoreMethod**, which provides the facilities and properties required by Chameleon. Most of the base class implements the mechanisms by which methods can communicate with each other, and be controlled by the system or user. All methods must be subclassed from this base method. Methods based on the same technique will be classed together in a new class, so for instance, there might be a class **TextMethod** which is a subclass of **CoreMethod** which includes text based methods. We envision that new classes would be created for problem or domain specific methods.

18.3.2 External methods

Although our design incorporates the ability to define fairly abstract methods, our major interest is the interactive graphical exploration of scientific problems. One of the basic requirements, then, is the provision of a range of display algorithms. There already exists a large body of software in this area, whether in the form of subroutine libraries, such as the NAG Graphics Library[2], or algorithms published individually, in *ACM Transactions on Mathematical Software*, for example. Rather than waste many hours re-inventing many wheels we hope to inherit these tried and tested algorithms. External methods have been designed to facilitate this process.

External methods will be based on a standard graphics package, such as GKS or PHIGS PLUS. Chameleon will be responsible for creating the workstation (GKS) or equivalent drawing area, which will then be passed to the method. The external software should make no attempt to provide any user interface, but should merely provide a display capability. Since many algorithms are already written in GKS, their conversion for use with Chameleon should be relatively painless. The method writer also has to provide a *method descriptor* which describes the properties of the method, giving their type, default values etc.

18.3.3 Implementation details

Chameleon has been implemented on the X Window System[10]. X has become the *de facto* standard window system, and its adoption by all major workstation vendors ensures portability. We have chosen to build on the Xt Intrinsics toolkit[5] and the Athena Widget Set[7]. The use of Intrinsics meant that much of the user's visual configuration could be handled via the Intrinsics' *resources* mechanism.

Each method is implemented as a widget. Widgets provide an object oriented abstraction of the components used to create a graphical interface. One constituent of the Chameleon system is therefore a consistent set of visualisation widgets. Although they have been designed with Chameleon in mind, they are available as a stand-alone widget set.

The MIT X Consortium has defined a set of rules which well-behaved application programs should adhere to in order to co-exist peacefully with other applications. The rules are contained in the Inter Client Communication Conventions Manual (ICCCM)[9]. Chameleon is designed to be ICCCM compliant.

18.4 The view concept

A Chameleon view incorporates an instance of a method and its user interface, the exact format of which is determined by the user. From the user's standpoint, the view is the basic module in Chameleon. The user is not limited in the number of views permitted on screen at any time, and is free to create, organise and destroy them at will. Views can be created from scratch, or can be spawned from existing views, inheriting their context.

The user interface for a method is constructed automatically from the method's descriptor, which as stated in the previous section, contains declarations of the method's properties. The user is presented with an appropriate interactor for viewing and modifying any property of the method. Chameleon holds the descriptors for currently available methods in a list, which can be queried by a view. In this way, each view is able to determine which methods can be spawned off, and presents a list to the user for selection. The contents of this list can be over-ridden in one of the profiles, for example to restrict the methods available from a particular view.

Chameleon contains a standard set of mechanisms for spawning new views. For example, given a contour plot, the user could define a sub region for a new contour plot, or select a line for a cross-section line graph. In addition each method can provide additional mechanisms to override the defaults. The ability to spawn new views from existing ones results in a hierarchy, with a tree of views for each problem.

18.4.1 View contents

Views are built around a common framework which serves to provide a standard 'look & feel' for the user to interact with different methods. This approach should reduce the learning time for new users and makes it easier to include additional methods. Associated with each method is a description of its properties which is used to build the interface around the method. A typical view might contain the following elements:

- an instance of a method.

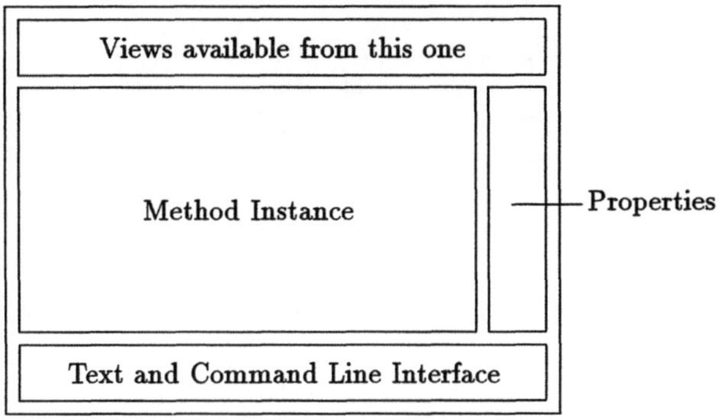

FIGURE 18.2. Major view components

- mechanisms for modifying properties of the method.
- a list of methods available for spawning off from this view.
- a text window to keep the user informed at all times.
- a standard help mechanism.
- a command line interface (CLI) to the view.

The default layout scheme for views is illustrated in figure 18.2, and an example view, containing a contour plot method, can be seen in figure 18.1.

18.5 User interface

The interface for a view can be presented in a number of different styles, with the user able to switch between them as the occasion demands. The interaction modes are not mutually exclusive, and can all be visible and active at the same time. These ideas are similar to those presented by Kantorowitz and Sudarsky for their *Adaptable User Interface*, where the dialogue modes are different representations of a single underlying *dialogue language*[4].

The inclusion of different interface paradigms is intended to make the system attractive and usable to as wide a range of users as possible. It is also an attempt to provide an interface which can grow with the user's experience and requirements. Consider three types of user:

1. *Casual user or beginner.* For this type of user, the most important considerations are often ease-of-use and gradient of learning curve. Full power and flexibility can be traded against an intuitive and uncluttered interface. The amount of text and user typing required should be kept to a minimum. The simplest interface in Chameleon uses icons to represent actions, attributes and other views, and attempts to keep the user's hand on the mouse as much as possible. It also has to be remembered that many users will base the decision of whether to use a system on the first few minutes of use, often running it with no previous knowledge.

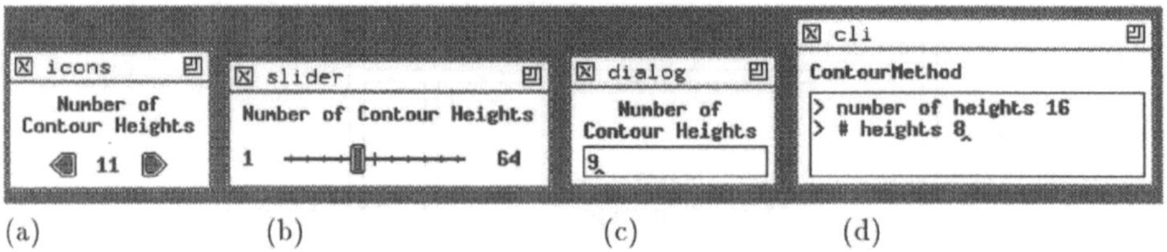

FIGURE 18.3. Techniques for modifying an integer property

2. *Domain specific user.* A user from a particular subject area, who may have provided additional, domain specific, methods. For these methods, he will want access to all features of the method, so will require a more powerful interface than the beginner. For other methods though, he may well prefer a simpler interface, which means he should be able to configure the interface either on a per-method basis, or a per-method-class basis.

3. *Experienced user.* An experienced user will want access to the full capabilities of any method, but may also want to save screen space by switching to a simple interface for specific views. For certain operations it may also be easier for user to type a command into a CLI.

As an example, the **SolidContour** method has a integer property which specifies the number of contour heights to be displayed. Figure 18.3 illustrates four ways in which this property could be modified. In (a) clicking the mouse button on the arrows to either side of the current value will increment or decrement the value. (a) and (b) remove any possibility of an illegal value being entered, but some users would find them restrictive. One advantage of the CLI is that only one interaction window is used to set any property, which is efficient in screen space, and lets the user's focus stay in one place. A number of techniques for entering values have been proposed, often with the intention of keeping the user's hand on the mouse, for example Ressler's *incrementor* which can be used to modify numeric parameters[8].

The user is not constrained to one interface style: a view can include any combination of the different approaches. This allows a smooth transition across paradigms as the user grows in experience and confidence. Having selected a particular style for a view, the user is not tied to it. If he has forgotten how to achieve something from the command line, or feels restricted by the iconic interface, then the interface could be changed, or another added. As users gain in experience, they can modify the working environment from within Chameleon and ask that the changes be reflected in their profile (discussed in the next section). Chameleon also provides a configuration/profile editor. This can be used to set a wide range of user preferences - either globally; for particular problems; for classes; or for named instances.

18.6 Problem interface

Another point which needs addressing is the ability to communicate with the problem solving code. This requires an interface between Chameleon and the problem code, which has two basic constructs:

1. Attributes, or variables. These are used to modify parameters of the computation. They could represent direct attributes of the computation, or, for example, specify a file containing new data. Referring to the Azeotropy problem, each chemical has a named file containing chemical data. The user could be presented with a list of known chemicals, from which he could select a number to mix together.

2. Commands. These would initiate some action in the problem code, for example to move to the next time step, or request remeshing. In the azeotropy example, after selecting the required liquids, the user could request they be loaded, and the new azeotropy function displayed.

The problem owner includes an interface in the problem code, and presents Chameleon with a *Problem Interface Descriptor*, from which Chameleon would construct additions to the existing interface. Note that although Chameleon controls the presentation aspects of the interface, the contents are dictated by the problem.

18.7 Help

An important aspect of any system is the documentation and other related information available to the user, and how it is accessed. Unfortunately, many systems require that the user reads a manual (or at least a tutorial to start off with) before running any program. Some users undoubtedly prefer this approach, but we must cater for those who prefer a 'hands on' approach.

It is useful to discuss the various types of help that might be needed at different times:

1. *Introductory.* This is intended for the beginner. It should include simple descriptions of what Chameleon can and can't do, and outlines of the major components and how they work.

2. *System.* High-level help about the system itself, or a particular component and how to get it going.

3. *Localised.* Information made available by the different components of the system. This might be of the form *'How do I do this'* or *'What does this button do'*.

There are three events which might result in help being provided: the user has requested it, the user has made a mistake, or the system thinks the user will benefit from it. The ability to provide useful help for all cases implies at least a limited form of user modelling, in order to try and determine what information the user is actually after. As a simple example, a user who is using only the simple iconic interface would not require the same help as someone using the full interface or a CLI.

Very often the help component of a system is seen as totally separate, with all information drawn from a different source. This information is usually not maintained with the rest of the system, resulting in inaccurate help being given. In Chameleon, the help subsystem uses, wherever possible, the same information as the rest of the components of Chameleon. The help itself is presented using the same user interface style that the user is using.

In figure 18.1, the user has requested help with a specific view by clicking on the '?' in the bottom right corner of the view. The view is now in help mode, which is reflected in the mouse pointer changing to a question mark. The user can now click on a component of the view and will be provided with help. Example help windows can be seen on the right-hand side of figure 18.1.

18.8 Configuration

This section outlines the various types of configuration available to the user. One inevitable consequence of providing a flexible and customisable system is that the need for configuration and profile files is introduced. We realise that these can be time consuming to edit, and could deter first time users, so anything which can be set in a configuration file has a built-in default. This means that Chameleon can be run without any configuration files.

18.8.1 System configuration

A *system configuration* is a description of the components requested in Chameleon. Two basic formats for Chameleon are available:

Data visualisation system, which can be used to view precalculated sets of data only.

Function and data visualisation system, where one or more user supplied functions are built in with the system.

The system configuration may include information such as a list of methods to include, system defaults, problem configuration(s), and extra resources, such as named colourmaps.

18.8.2 Problem configuration

A *problem configuration* describes a single user problem, defined either by a function or a set of data. This file provides the following information:

- details about the function: the function itself, an optional user-supplied initialisation routine, any known values of interest (such as minimum or maximum) etc.
- details about the data set: No. of variables, size, file name, format, minimum & maximum etc.
- settings for any properties.
- list of methods required, and an initial method.
- additional imagery for use in the interface.
- any problem specific help and other textual information.

18.8.3 User profile

The *user profile* is used to supply user preferences and additional information on a per-user basis. The information provided could be:

- aesthetics: colours to use, reverse video etc
- working modes: global, per-class or per-method.
- default methods
- user directories with additional information for Chameleon: data files, colourmaps, destination for logging & other output.

18.9 Chameleon

The previous sections of this paper introduced the major components and concepts in Chameleon. In this section we will briefly outline how the system is put together and how someone goes about using it.

18.9.1 Starting Chameleon

By the time the user is ready to run Chameleon, he should have provided one or more problem definitions, either by linking them in, or providing precalculated data. When Chameleon is running, each problem is given a context which holds run-time information on the current state of the problem. On running Chameleon, the Context Manager will appear, with an entry for each problem defined. A default problem can be specified, which becomes active on startup.

By default, when Chameleon starts up, only the context manager is created. If users find that for certain problems there are methods which they always use, then they can modify the problem's configuration to request that views containing the methods be created on startup. The configuration can either be modified by hand, using a standard text editor, or the user can ask that the current state of a problem be reflected in its configuration.

18.9.2 Using Chameleon

Chameleon incorporates a simple version of the Rooms concept introduced by Henderson & Card[3]. On startup, each problem is assigned a separate *room* which will contain all views associated with the problem. Whilst standing in one room, the views in another room are not visible. Switching contexts (ie moving rooms) results in the views from the original context being iconified, and the views for the new context being opened.

Each problem gives rise to a hierarchy of views, which can be viewed graphically as a tree from within the context manager. Attributes of views can be modified from this overview, for instance making a view visible in other rooms.

The user can modify attributes of views and their user interfaces, saving the modifications either to the user's profile, or a named file. The user can also request a 'snapshot' of the system, so that the current state can be recreated at a later date.

18.10 Conclusions

18.10.1 Concluding remarks

In this paper we have presented our ideas for an extensible scientific visualisation system. Although the specification is mostly complete, we have currently implemented only prototype methods and views, to help clarify our ideas.

18.10.2 Future work

Implement Chameleon! We also have a number of ideas for possible extensions and future work on Chameleon:

- Improve the mechanism for including the scientific computation. This is very simplistic at the moment.
- Include a mechanism by which the user can define the user interface to a greater extent. Users should be able to interact with the routines that they have added.

- We need to consider the visualisation of problems which include time as one of their independent variables. Although methods for this type of problem could be implemented under the current design, we have to decide whether Chameleon could be extended to facilitate their design.
- The standard set of methods in Chameleon should be extended to include generic techniques for displaying multi-dimensional data. Bergeron & Grinstein have proposed a reference model for the visualisation of multi-dimensional data, which includes generic techniques[1]

Acknowledgements:

N. Bowers is supported by SERC and NAG Ltd through a CASE award.

18.11 References

[1] R.D. Bergeron and G.G. Grinstein. A Reference Model for the Visualization of Multi-dimensional Data. In *Eurographics '89*, pages 393–399, 1989.

[2] K.W. Brodlie et al. The Development of the NAG Graphical Supplement. *Computer Graphics Forum*, 1(3):133–142, September 1982.

[3] Jr Henderson, D.A. and S.K. Card. Rooms: The Use of Multiple Virtual Workstations to Reduce Space Contention in a Window-Based Graphical User Interface. *ACM Transactions on Graphics*, 5(3):211–243, July 1986.

[4] E. Kantorowitz and O. Sudarsky. The Adaptable User Interface. *Communications of the ACM*, 32(11):1352–1358, 1989.

[5] J. McCormack, P. Asente, and R.R. Swick. *X Toolkit Intrinsics - C Language Interface.* MIT X Consortium.

[6] B.H. McCormick et al. Visualisation in Scientific Computing. *Computer Graphics*, 21(6), November 1987.

[7] Chris D. Peterson. *Athena Widget Set - C Language Interface.* MIT X Consortium.

[8] S. Ressler. The Incrementor: A Graphical Technique for Manipulating Parameters. *ACM Transactions on Graphics*, 6(1):74–78, January 1987.

[9] David S.H. Rosenthal. *Inter-Client Communication Conventions Manual.* Sun Microsystems, Inc.

[10] R.W. Scheifler and J. Gettys. The X Window System. *ACM Transactions on Graphics*, 5(2):79–109, April 1986.

Colour Plates

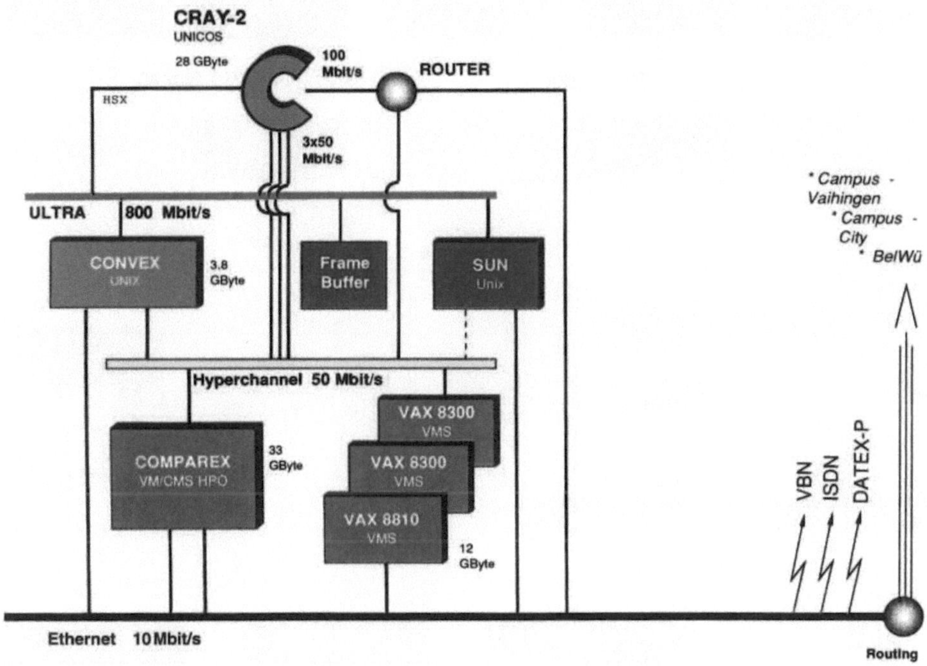

Plate 1. RUS computer configuration

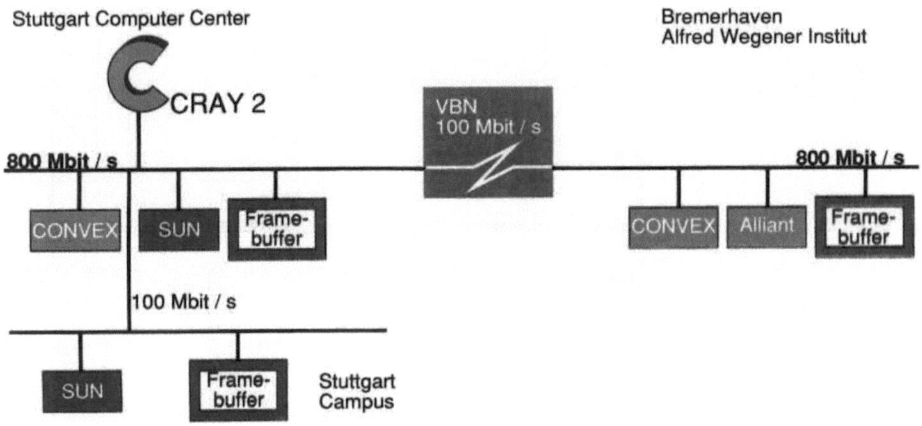

Plate 2. Ultranet high speed connection

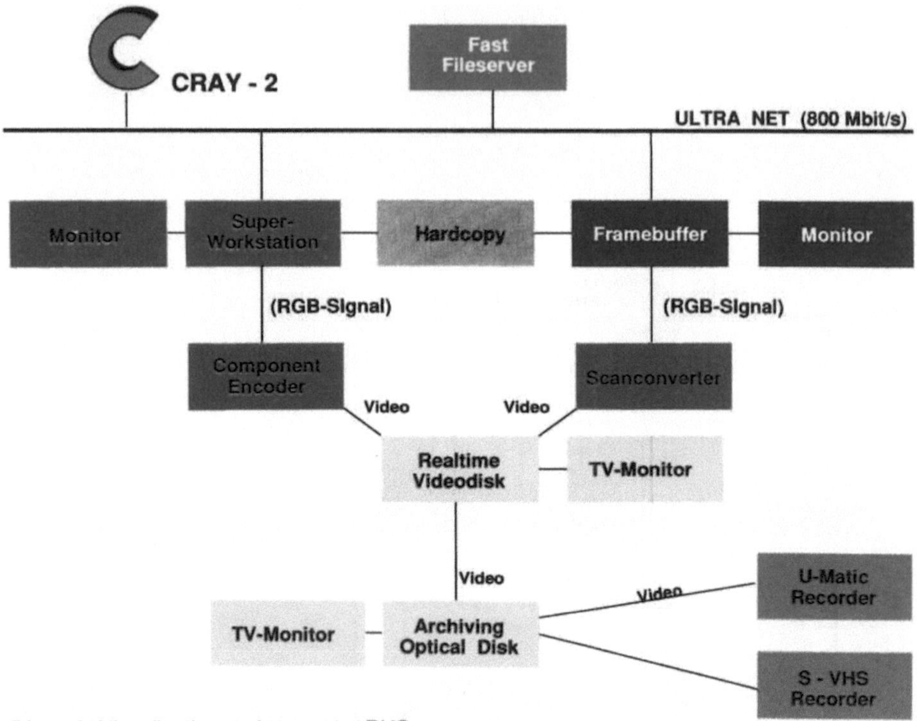

Plate 3. Visualization environment at RUS

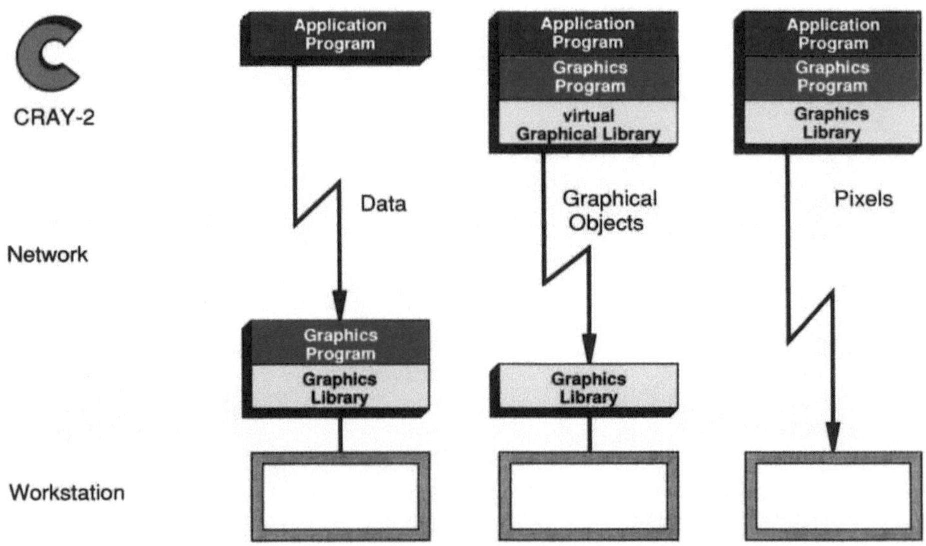

Plate 4. Information exchange in visualization applications

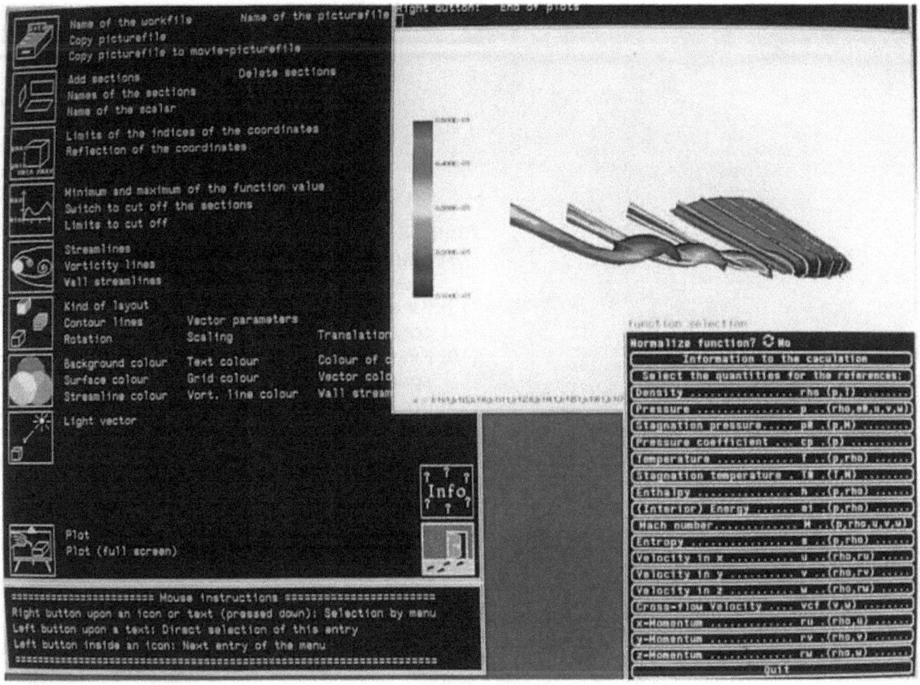

Plate 5. User interface for calculating additional quantities

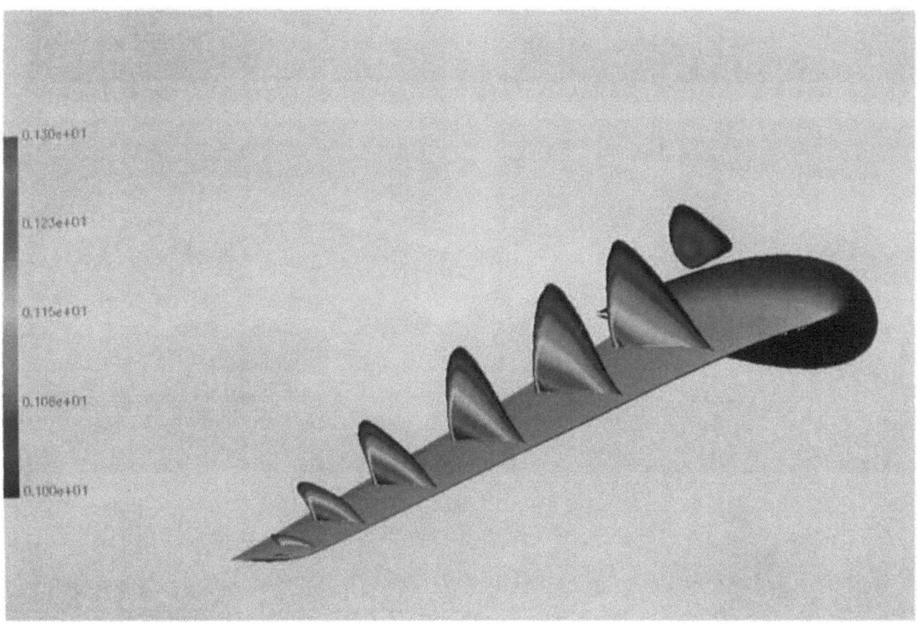

Plate 6. Mach number distribution inside supersonic flow region

Plate 7. Streamlines in a turbine flow

Plates 8–11
(8) Kinetic energy and water vapour specific humidity at hour 113
(9) Kinetic energy and water vapour specific humidity at hour 125
(10) Kinetic energy and water vapour specific humidity at hour 135
(11) Potential temperature at hour 120

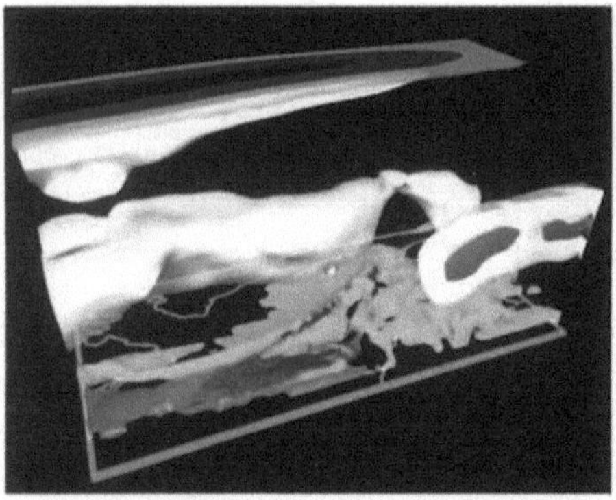

8

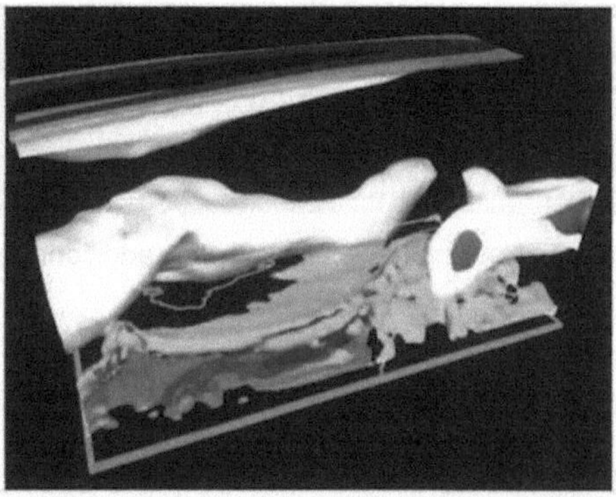

9

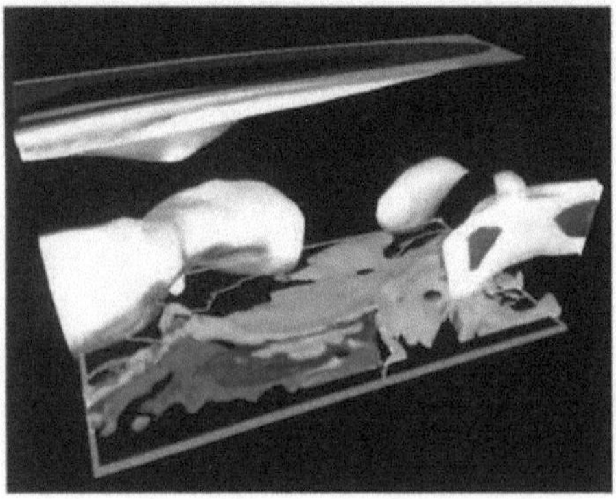

10

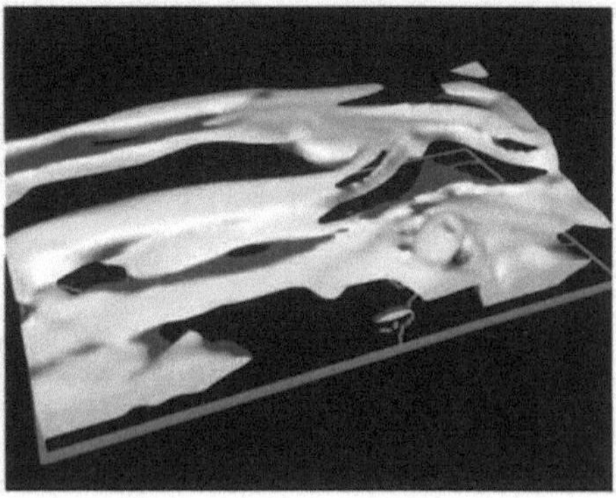

11

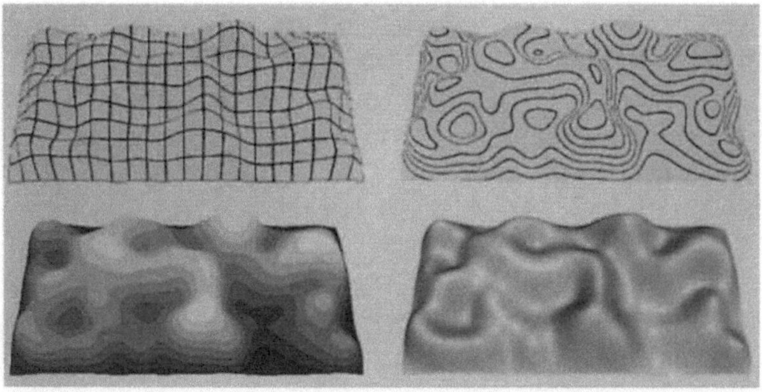

Plate 12. Four representations of a two-dimensional function:
(b) contour lines,
(c) grey shades indicate height,
(d) light reflection

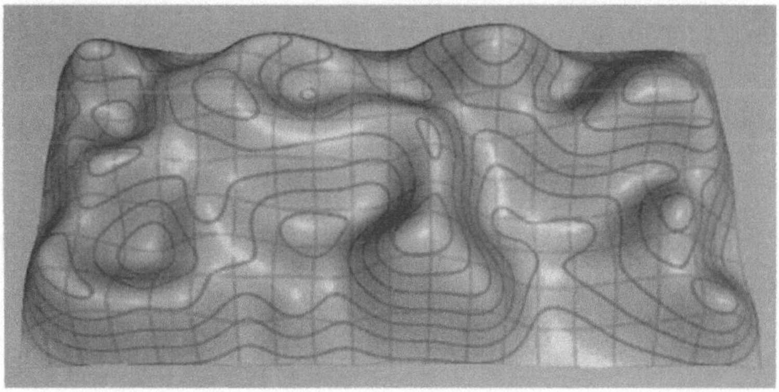

Plate 13. Contour lines and grid lines over a shaded surface

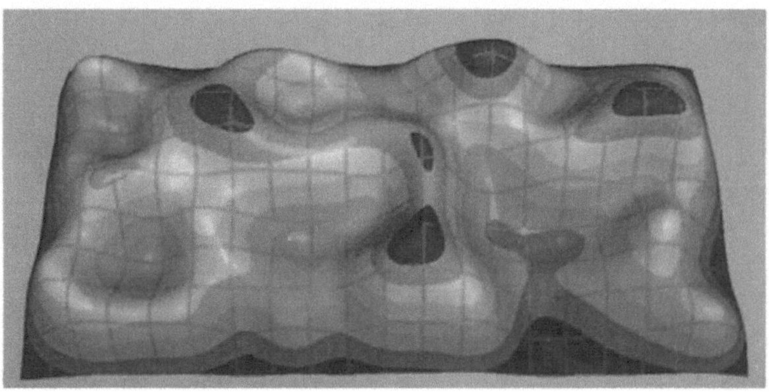

Plate 14. Pseudo-colours and grid lines over a shaded surface

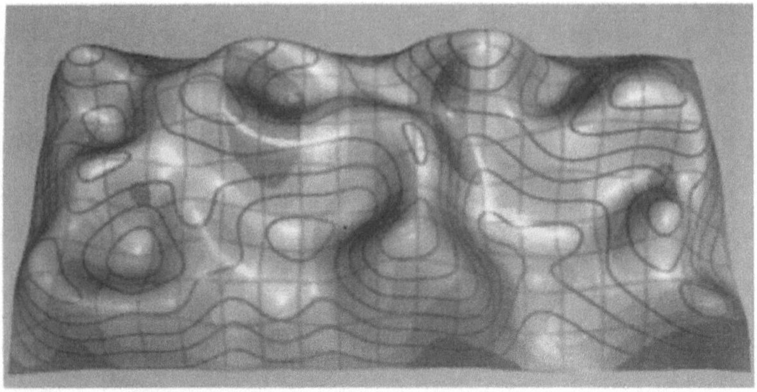

Plate 15. Pseudo-colours representing a second function

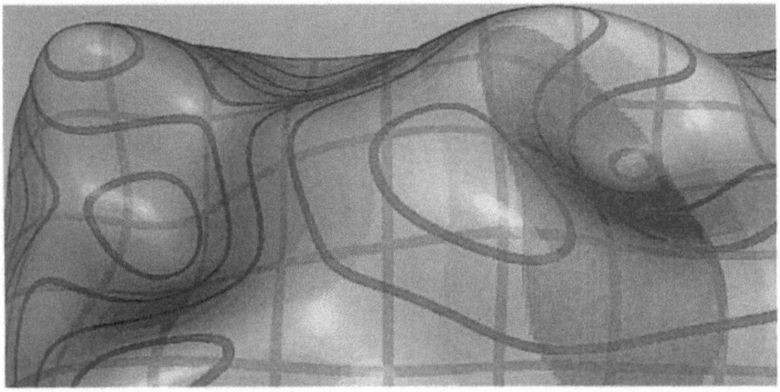

Plate 16. Close-up view of plate 15

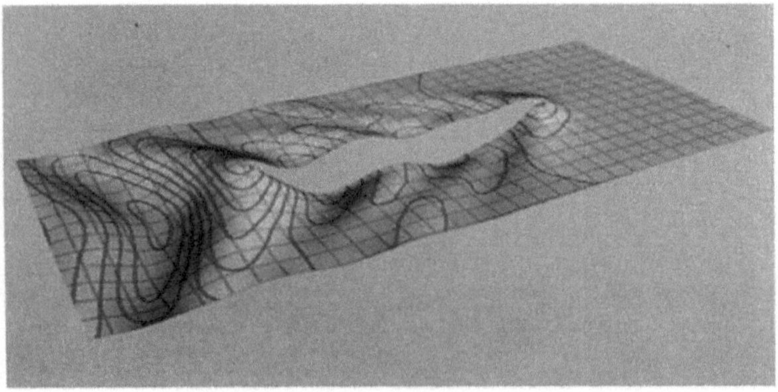

Plate 17. Wave pattern of a ship (courtesy data: H. C. Raven, Maritime Research)

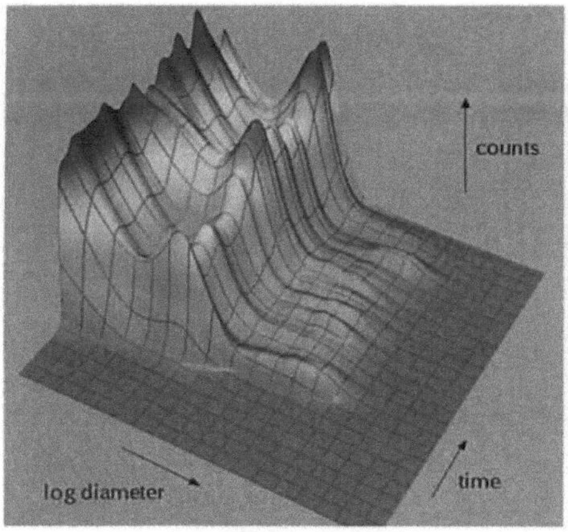

Plate 18. Droplet distribution of fog (courtesy data: B. G. Arends, The Netherlands)

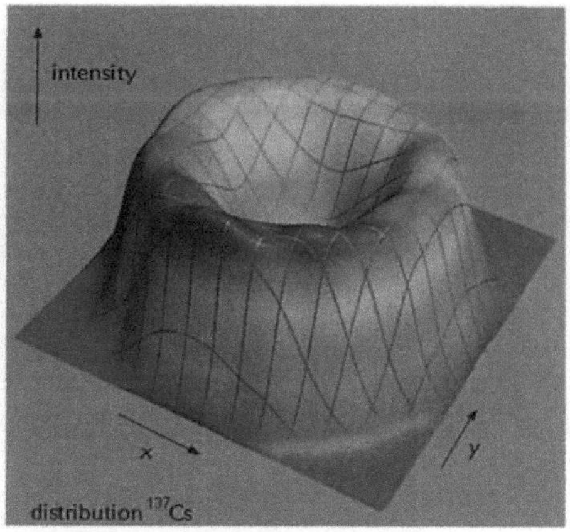

Plate 19. Distribution of ^{137}Cs nuclide over fuel rod

20

21

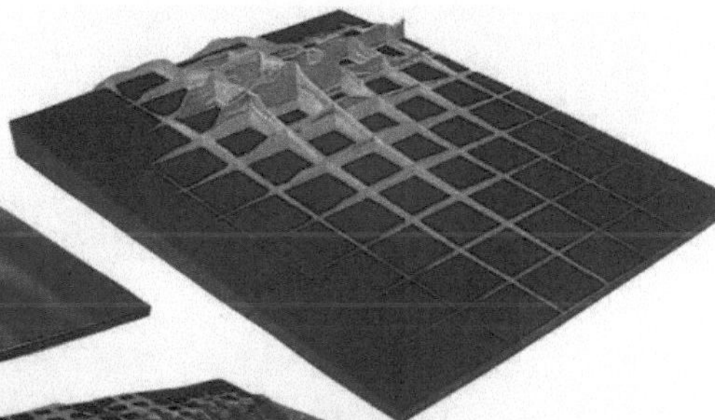

22

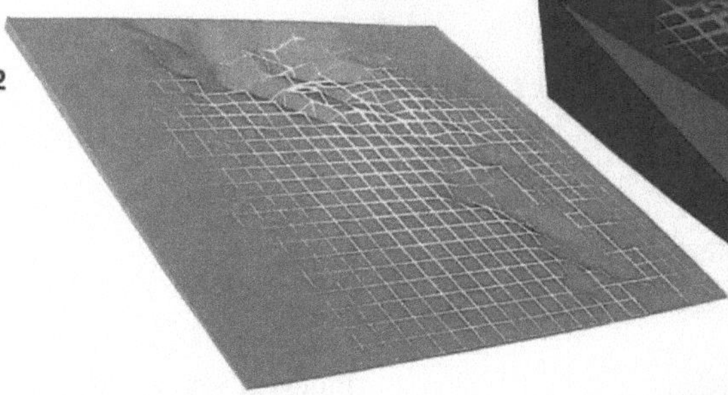

23

24

Plates 20–24

(20) Isochron sediment surface
(21) Fence diagram
(22) Basement topography and fence diagram, 2-fold exaggerated; note erosion of basement
(23) Sedimentary basin: A river carrying sediment enters at the top. Wave activity segregates sand (red and orange) from finer material (green and blue), driving sand to the right parallel to shore. Note the shoreline bounding the water body
(24) Sedimentary basin: Deposit surface and interior can be viewed simultaneously if the surface is rendered translucent. A set of graphic controls lets the user interact with the display

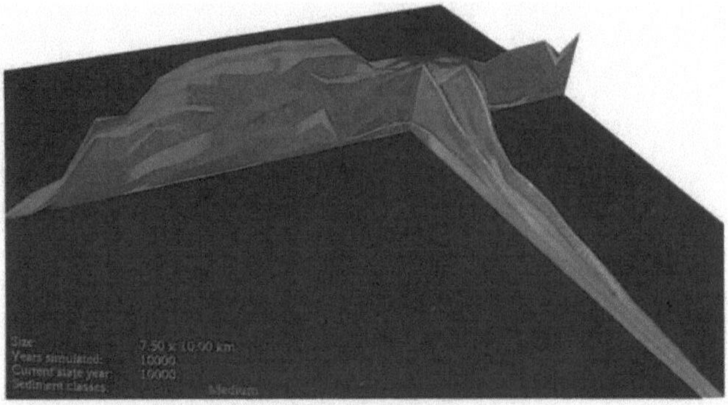

25

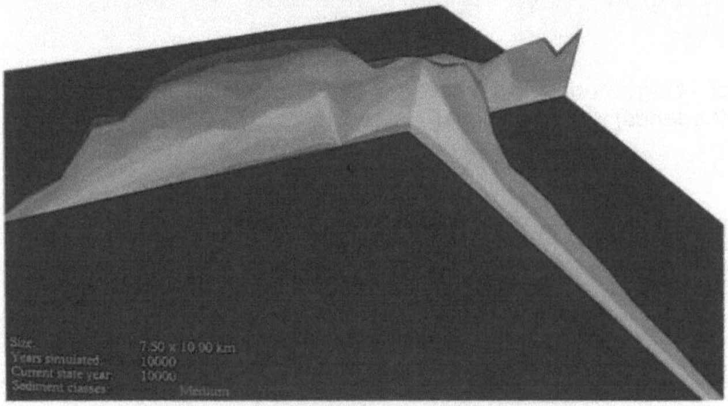

26

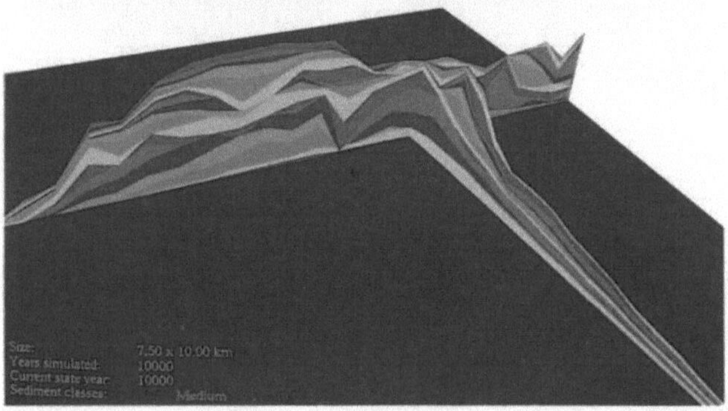

27

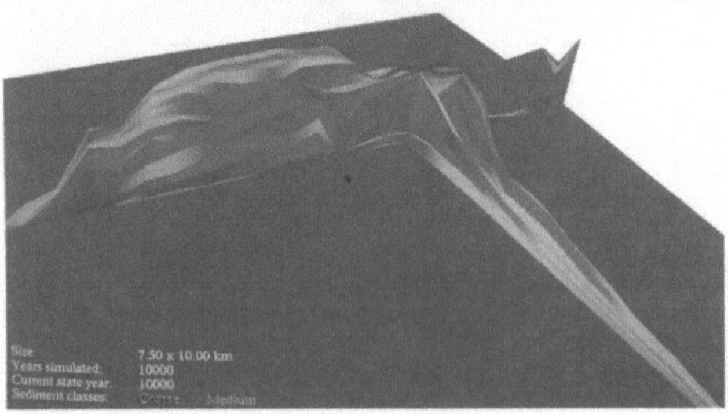

28

Plates 25–28

(25) Sediment classification by sediment type

(26) Sediment classification by sediment age. Smooth colour transitions enhance discontinuities

(27) Sediment classification by sediment age. Distinct colours enhance layer boundaries

(28) Sediment classification – highlighting of medium grain size

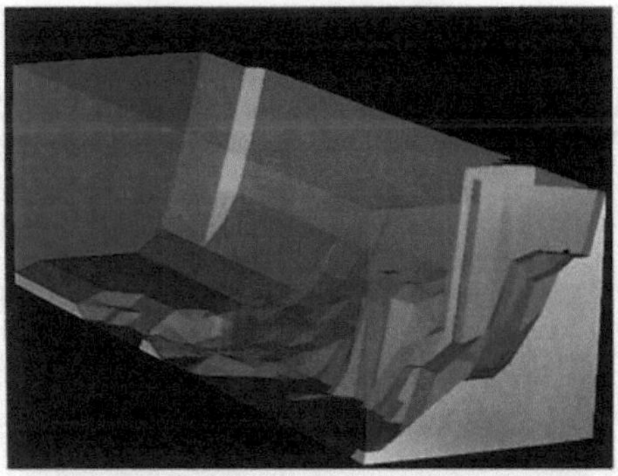
29

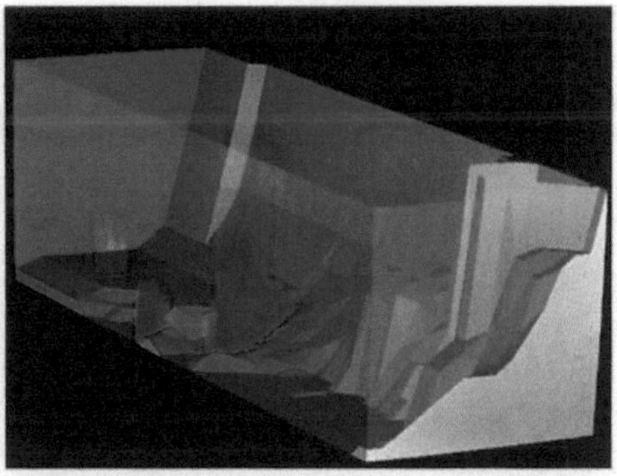
30

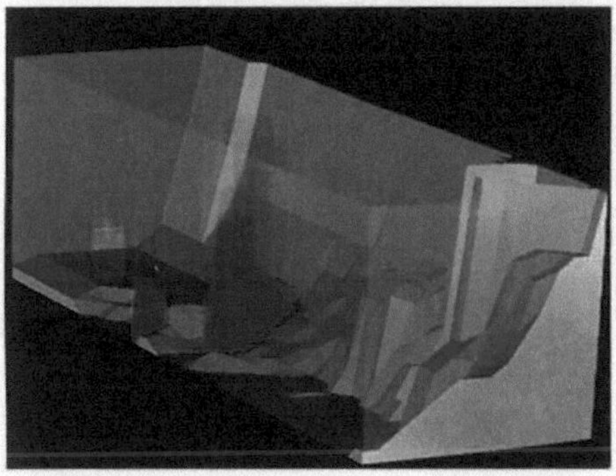
31

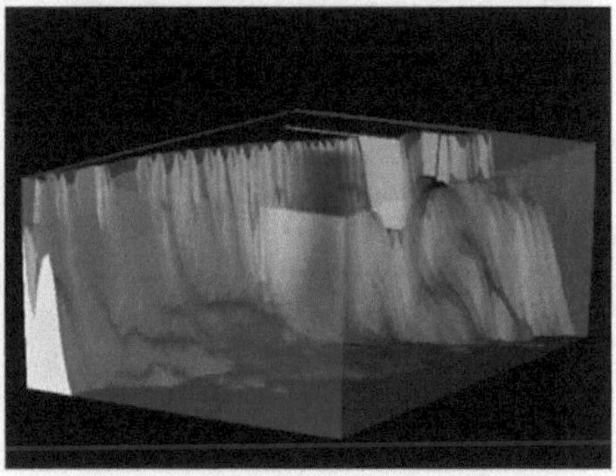
32

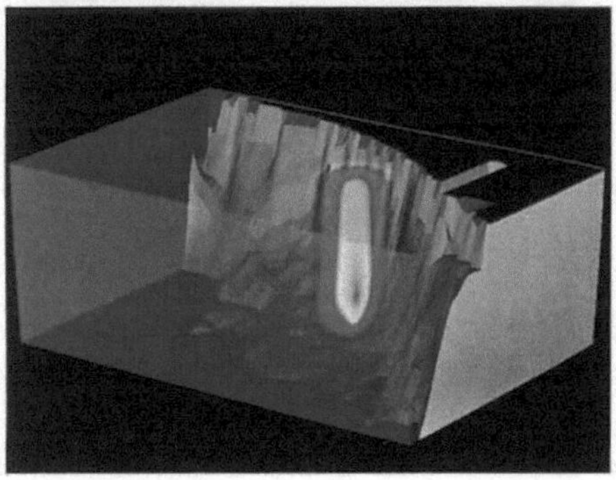
33

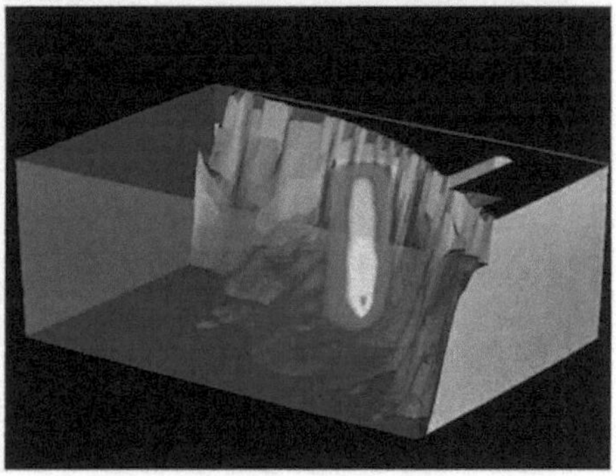
34

Plates 29–34

(29) Aquarium model with trilinear interpolation in cells, compression factor = 1
(30) Trilinear interpolation in cells, compression factor = 5, depth cueing on bottom
(31) Trilinear interpolation in cells, compression factor = 10
(32) Aquarium model with trilinear interpolation in cells, compression factor = 1
(33) Constant-valued cells, compression factor = 1, colour range: red for maximum values via green and blue to magenta for zero values
(34) Trilinear interpolation in cells, compression factor = 1, the same colour range as in (33)

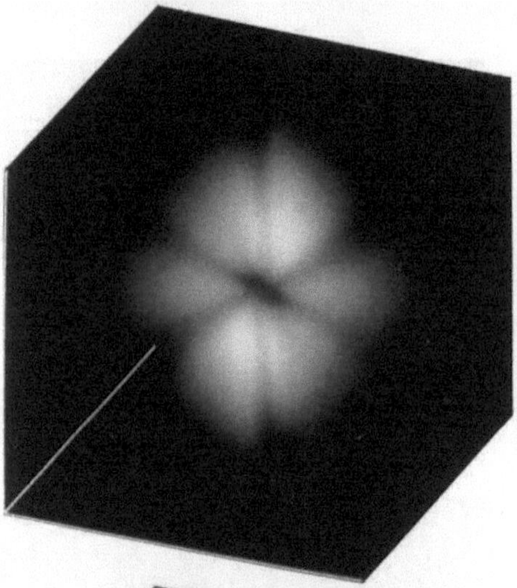

35

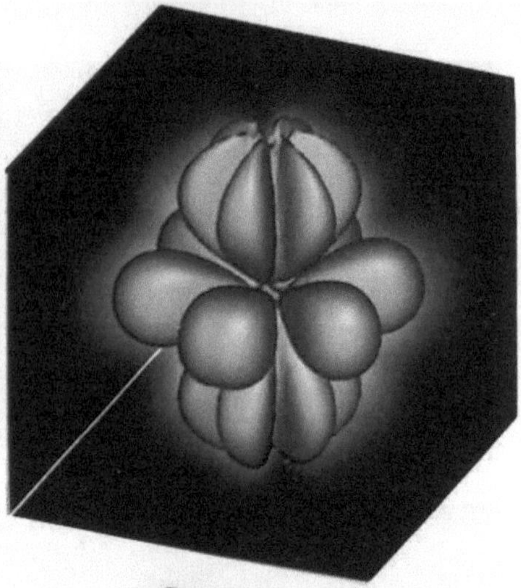

36

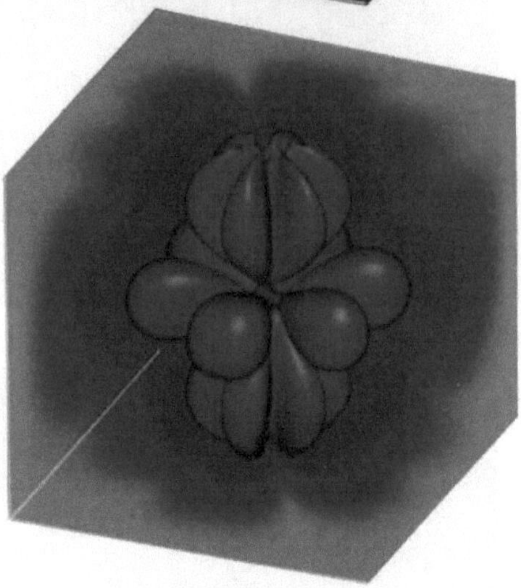

37

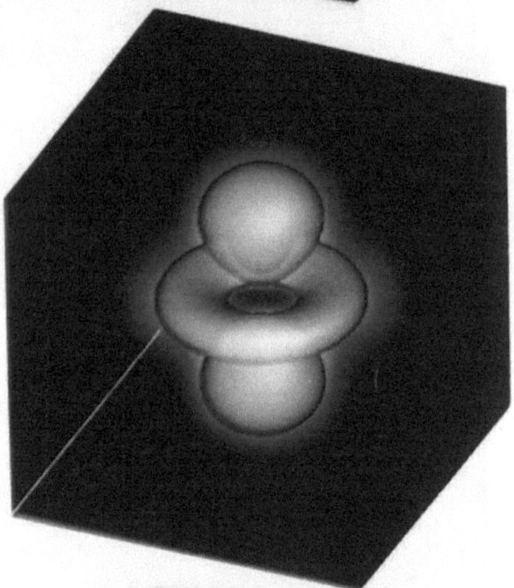

38

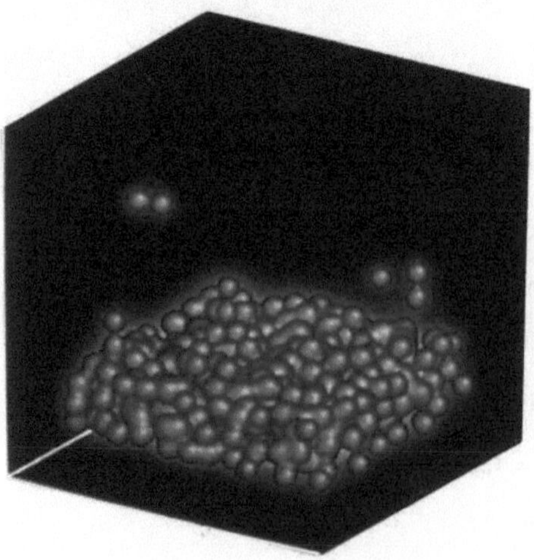

39

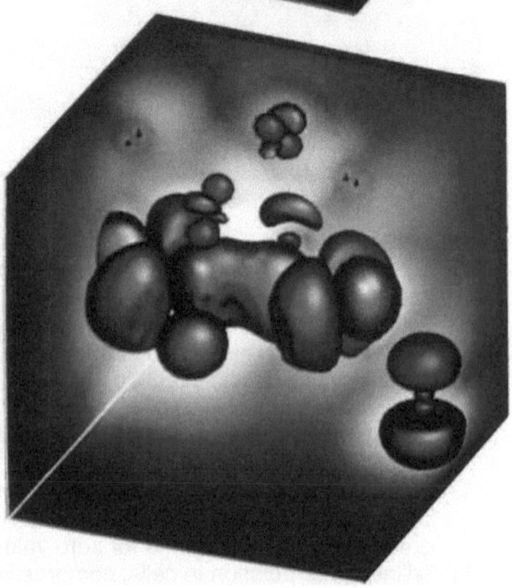

40

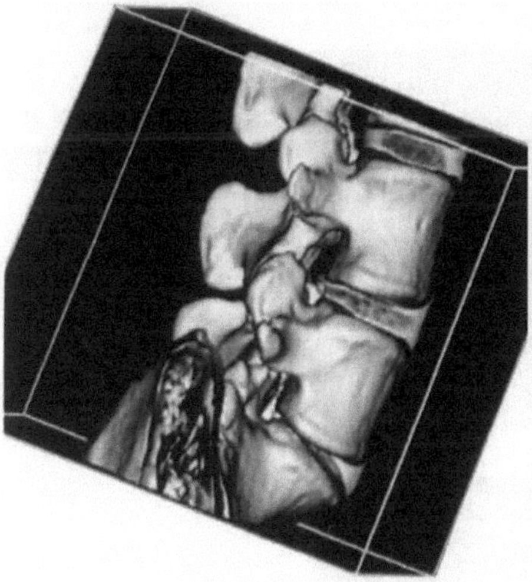

41

42

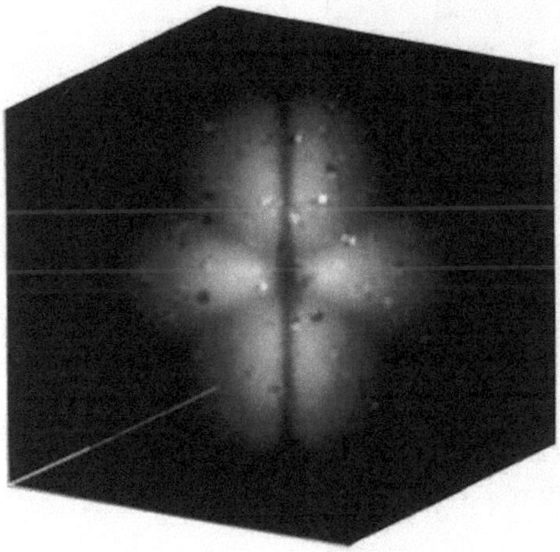

43

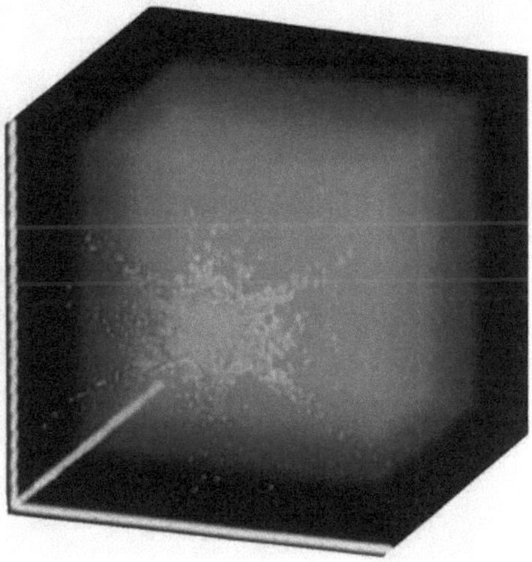

44

Plates 35–44

(35) Action of the pure volume term q_v (Ψ-function for the electron in a highly excited H-atom)

(36) Role of a surface-like source term q_s (same data field)

(37) Role of the volume absorption term (same data field)

(38) Role of specular term and transparency for enhanced depth information (electron density for the same H-atom)

(39) Role of vibrating atoms of a crystal lattice

(40) Role of the colour shift term S_{in} (Ψ-function of an iron protein, data field provided by L. Noodleman and D. Green, Scripps Clinic, La Jolla, CA)

(41) Visualization of a medical CT-data set by combining mappings onto surface and volume source terms (data provided by A. Kern, Radiological Institute of the FU Berlin)

(42) Of the "distance" of two related data fields on a isosurface using random texturing

(43) Of point-like deviations of two related data fields via the volume source model using random texturing

(44) Pattern for a lattice with diamond-like symmetry

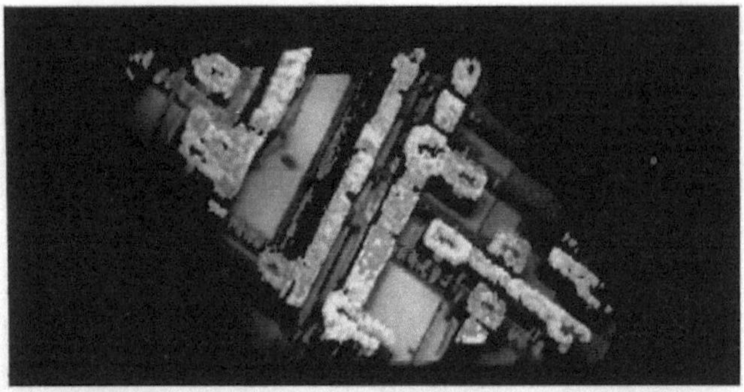

45

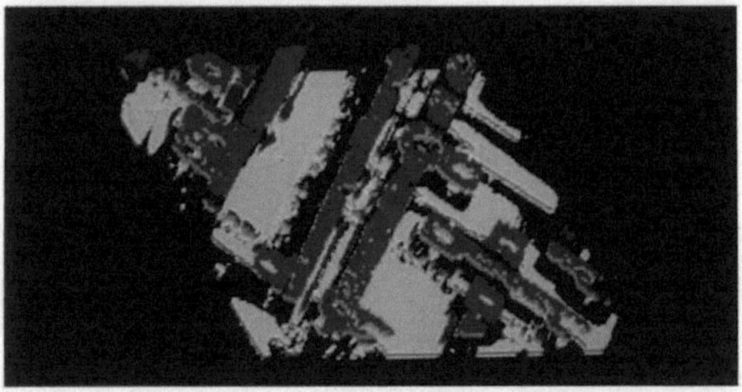

46

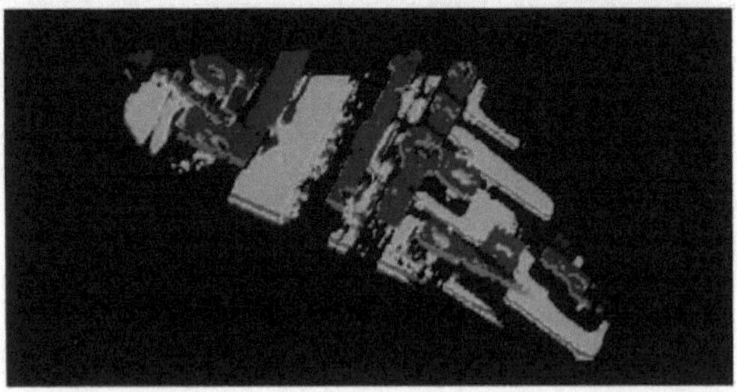

47

Plates 45–47
(45) View Mode
(CLSM Image of an Integrated Circuit, Resolution: 256*256*32)
(46) Layer Mode
(47) Cross-section

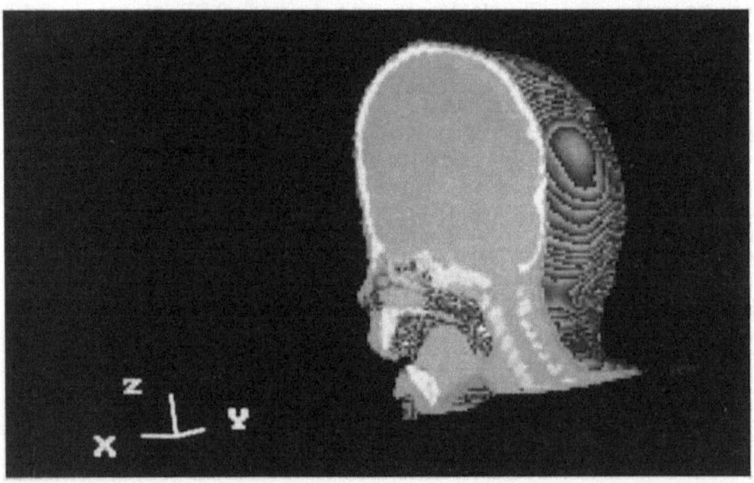

Plate 48. Volume of Interest. (CT scan of baby head, Resolution: 128*128*128)

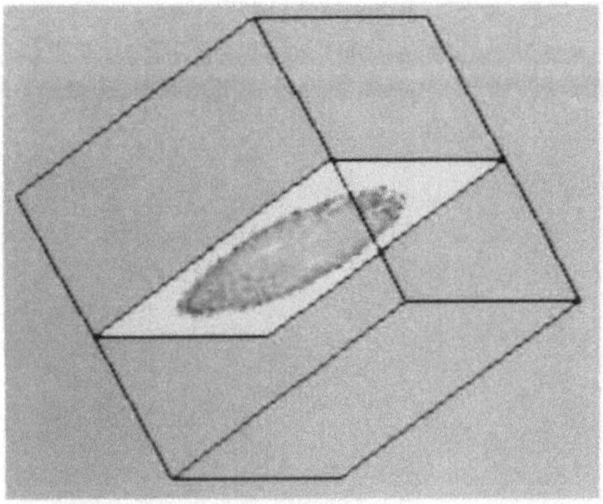

49

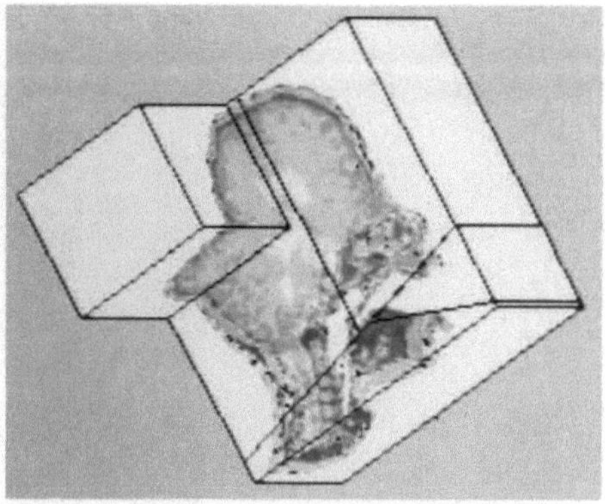

50

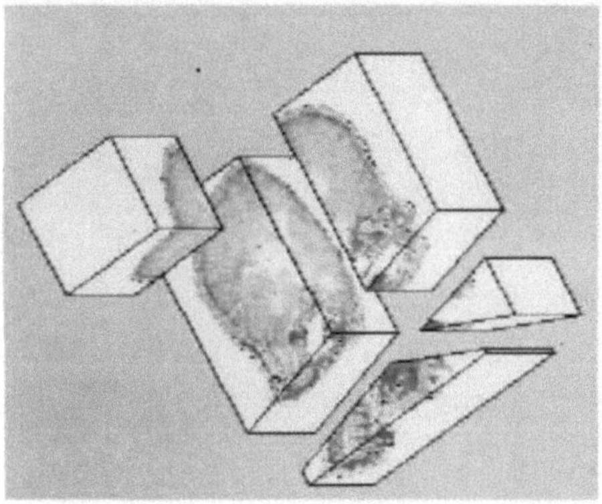

51

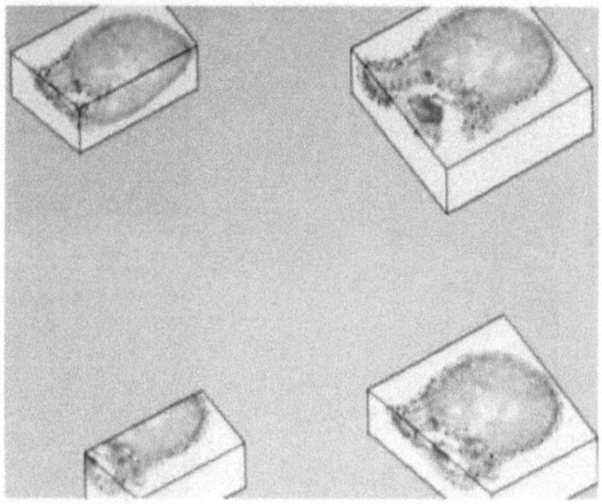

52

Plates 49–52[1]

The volume dataset in these plates consists of a voxel model of 128 3 8-bit voxels. The data were obtained from 128 CT scans. Original CT scans were images of 256 2 12-bit pixels. Reduction to 128 2 slices was done by averaging 12-bit pixel values over 2 x 2 pixel neighbourhoods and taking the 8 most significant bits of the resulting values. The colours shown do not have any clinical significance, but approximately show bone (green and blue), skin and subcutanuous fat (yellow), soft tissue (red), and air (white).

(49) Initial position of the first partitioning plane w.r.t the voxel cube
(50) Example of a subdivision. Some cells have been made invisible
(51) Subdivision shown in exploded view
(52) Part of the BSP-tree: current cell (top row, right), its parent cell (top row, left), and its two children (bottom row)

---

[1] All data courtesy of:
J. C. van der Meulen, Department of Plastic and Reconstructive Surgery, Rotterdam University Hospital Dijkzicht, Rotterdam, The Netherlands, F. W. Zonneveld, Department of Diagnostic Radiology, Utrecht University Hospital, Utrecht, The Netherlands and S. Lobregt, Philips Medical Systems, CT Sanner Science Department, Best, The Netherlands

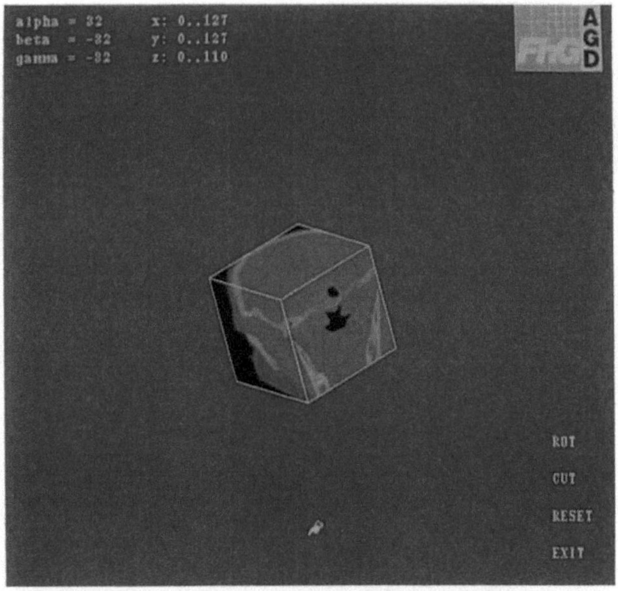
alpha = 32 x: 0..127
beta = -32 y: 0..127
gamma = -32 z: 0..110
ROT
CUT
RESET
EXIT

53 a

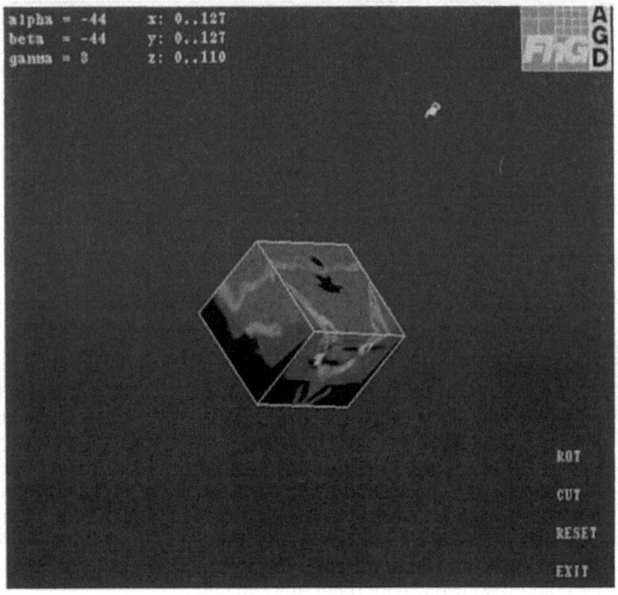
alpha = -44 x: 0..127
beta = -44 y: 0..127
gamma = 3 z: 0..110
ROT
CUT
RESET
EXIT

53 b

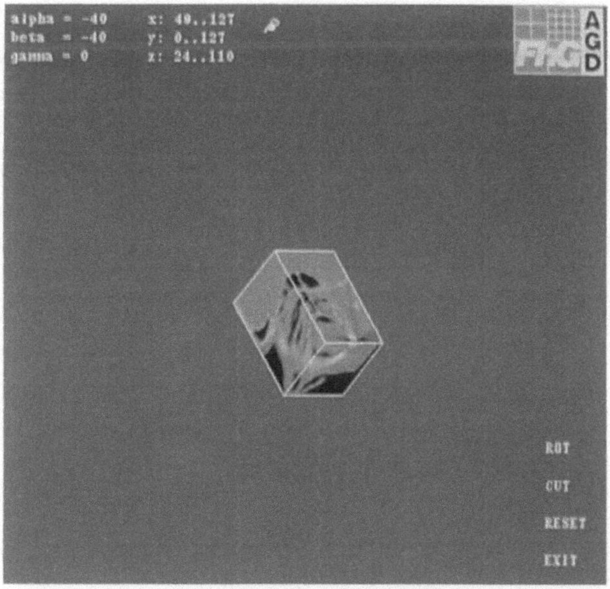

54

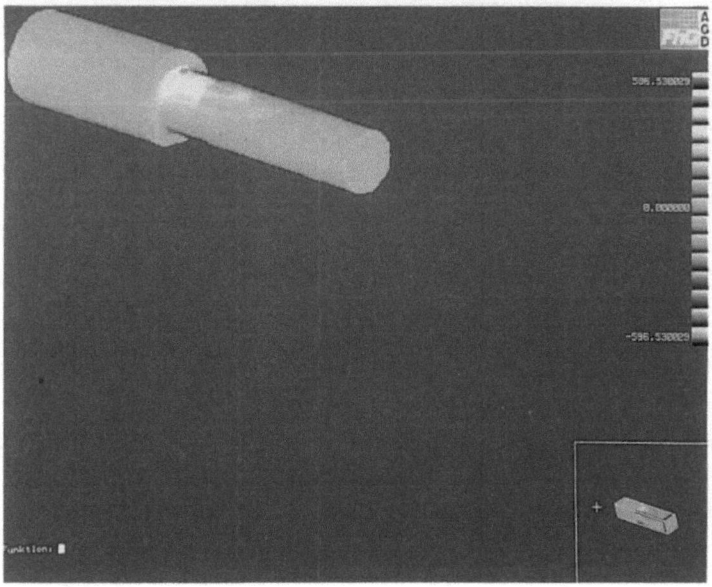

55

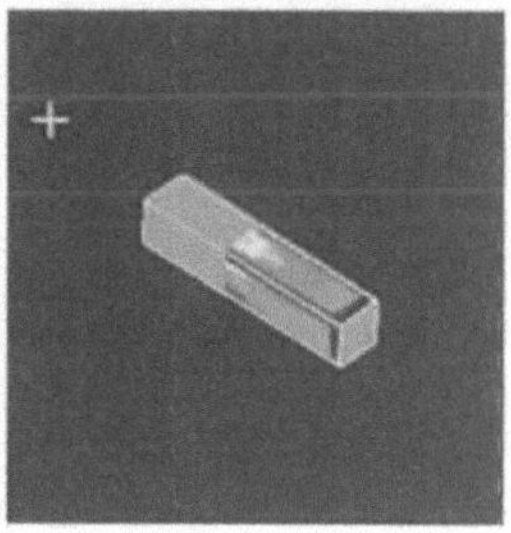

56

Plates 53–56

(53) Rotation of a CT data set (128 x 128 x 111 voxel)
(54) Cutting of the CT data set
(55) Selection of view point and volume rendering of finite element data set
(56) Rotation of a finite element data set with the object projected onto the faces of the volume

List of Authors

Dölf Aemmer
ETH
Integrated Systems Laboratory
Gloriastrasse 35
CH-8092 Zürich
Switzerland

H. Aichele
Universität Stuttgart
Allmandring 30
D-70569 Stuttgart
Germany

D. Beaucourt
EDF/DER, Service IMA
1 Avenue du Général de Gaulle
F-92141 Clamart Cedex
France

Edwin Boender
Delft University of Technology
Faculty of Math & Informatics
Julianalaan 132
NL-2628 BL Delft
The Netherlands

N. Bowers
School of Computer Studies
University of Leeds
Leeds LS2 9JT
United Kingdom

K. W. Brodlie
School of Computer Studies
University of Leeds
Leeds LS2 9JT
United Kingdom

Lesley Carpenter
NAG
Wilkinson House
Jordan Hill Road
Oxford OX2 8DR
United Kingdom

Philip C. Chen
Jet Propulsion Laboratory
Mail Stop 510–512
4800 Oak Grove Drive
Pasadena, CA 91109
USA

W. Felger
FhG-AGD
Wilhelminenstrasse 7
D-64283 Darmstadt
Germany

Martin Frühauf
FhG-AGD
Wilhelminenstrasse 7
D-64283 Darmstadt
Germany

R. Gnatz
Technische Universität München
Institut für Informatik
Arcisstrasse 21
D-80333 München
Germany

M. Göbel
FhG-AGD
Wilhelminenstrasse 7
D-64283 Darmstadt
Germany

Michel Grave
ONERA, DMI/CC
29 Avenue de la Division Leclerc
F-92322 Chatillon
France

P. Hemmerich
EDF/DER, Service IMA
1 Avenue du Général de Gaulle
F-92141 Clamart Cedex
France

Andrea J. S. Hin
Delft University of Technology
Faculty of Math & Informatics
Julianalaan 132
NL-2628 BL Delft
The Netherlands

Nancy Hitschfeld
ETH
Integrated Systems Laboratory
Gloriastrasse 35
CH-8092 Zürich
Switzerland

G. R. Hofmann
FhG-AGD
Wilhelminenstrasse 7
D-64283 Darmstadt
Germany

R. J. Hubbold
Department of Computer Science
University of Manchester
Oxford Road
Manchester M13 9PL
United Kingdom

D. P. Huijsmans
Rijksuniversiteit Leiden
Faculteit der
Wiskunde en Natuurkunde
PO Box 9512
NL-2300 RA Leiden
The Netherlands

W. Huiskamp
TNO Physics & Electronic Lab.
Parallel Processing Group
PO Box 96864
NL-2509 La Haye
The Netherlands

G. J. Jense
FEL-TNO
PO Box 96864
NL-2509 JG 's-Gravenhage
The Netherlands

Kennet Karlsson
FhG-AGD
Wilhelminenstrasse 7
D-64283 Darmstadt
Germany

Herbert Klein
Stanford University
Geophysics Department
Stanford, CA 94305-2225
USA

Wolfgang Krueger
ART + COM
Hardenbergplatz 2
D-10623 Berlin
Germany

Peter Lamb
ETH
Integrated Systems Laboratory
Gloriastrasse 35
CH-8092 Zürich
Switzerland

U. Lang
Universität Stuttgart
Allmandring 30
D-70569 Stuttgart
Germany

A. A. J. Langenkamp
TNO Physics & Electronics Lab.
Parallel Processing Group
PO Box 96864
NL-2509 La Haye
The Netherlands

Yvon Le Lous
EDF/DER, Service IMA
1 Avenue du Général de Gaulle
F-92141 Clamart Cedex
France

Rick Ottolini
Stanford University
Geophysics Department
Stanford, CA 94305-2225
USA

Hans-Georg Pagendarm
DLR
Bunsenstrasse 10
D-37073 Göttingen
Germany

H. Pöhlmann
Universität Stuttgart
Allmandring 30
D-70569 Stuttgart
Germany

Frits H. Post
Delft University of Technology
Faculty of Math & Informatics
Julianalaan 132
NL-2628 BL Delft
The Netherlands

Christoph Ramshorn
Universität Freiburg
Geologisches Institut
Albertstrasse 23-B
D-79104 Freiburg i. Br.
Germany

R. Rühle
Universität Stuttgart
Allmandring 30
D-70569 Stuttgart
Germany

P. L. J. van Lieshout
TNO
Physics & Electronics Lab.
Parallel Processing Group
PO Box 96864
NL-2509 La Haye
The Netherlands

Jarke J. Van Wijk
Netherlands Energy Research
Foundation
PO Box 1
NL-1755 ZG Petten
The Netherlands

Carlo E. Vandoni
CERN
Data Handling Division
CH-1211 Genève 23
Switzerland

Hanspeter Wacht
ETH
Integrated Systems Laboratory
Gloriastrasse 35
CH-8092 Zürich
Switzerland

Focus on Computer Graphics

(Formerly EurographicSeminars)

Eurographics Tutorials '83. Edited by P. J. W. ten Hagen.
XI, 425 pages, 164 figs., 1984. Out of print

User Interface Management Systems. Edited by G. E. Pfaff.
XII, 224 pages, 65 figs., 1985. Out of print (see below, Duce et al. 1991)

Methodology of Window Management. Edited by F. R. A. Hopgood, D. A. Duce, E. V. C. Fielding, K. Robinson, A. S. Williams. XV, 250 pages, 41 figs., 1985. Out of print

Data Structures for Raster Graphics. Edited by L. R. A. Kessener, F. J. Peters, M. L. P. van Lierop. VII, 201 pages, 80 figs., 1986

Advances in Computer Graphics I. Edited by G. Enderle, M. Grave, F. Lillehagen. XII, 512 pages, 168 figs., 1986

Advances in Computer Graphics II. Edited by F. R. A. Hopgood, R. J. Hubbold, D. A. Duce. X, 186 pages, 96 figs., 1986

Advances in Computer Graphics Hardware I. Edited by W. Straßer.
X, 147 pages, 76 figs., 1987

GKS Theory and Practice. Edited by P. R. Bono, I. Herman.
X, 316 pages, 92 figs., 1987. Out of print

Intelligent CAD Systems I. Theoretical and Methodological Aspects.
Edited by P. J. W. ten Hagen, T. Tomiyama. XIV, 360 pages, 119 figs., 1987

Advances in Computer Graphics III. Edited by M. M. de Ruiter.
IX, 323 pages, 247 figs., 1988

Advances in Computer Graphics Hardware II. Edited by A. A. M. Kuijk, W. Straßer. VIII, 258 pages, 99 figs., 1988

CGM in the Real World. Edited by A. M. Mumford, M. W. Skall.
VIII, 288 pages, 23 figs., 1988. Out of print

Intelligent CAD Systems II. Implementational Issues. Edited by V. Akman, P. J. W. ten Hagen, P. J. Veerkamp. X, 324 pages, 114 figs., 1989

Advances in Computer Graphics IV. Edited by W. T. Hewitt, M. Grave, M. Roch. XVI, 248 pages, 138 figs., 1991

Advances in Computer Graphics V. Edited by W. Purgathofer, J. Schönhut.
VIII, 223 pages, 101 figs., 1989

User Interface Management and Design. Edited by D. A. Duce, M. R. Gomes, F. R. A. Hopgood, J. R. Lee. VIII, 324 pages, 117 figs., 1991

Advances in Computer Graphics Hardware III. Edited by A. A. M. Kuijk. VIII, 214 pages, 88 figs., 1991

Advances in Object-Oriented Graphics I. Edited by E. H. Blake, P. Wisskirchen. X, 218 pages, 74 figs., 1991

Advances in Computer Graphics Hardware IV. Edited by R. L. Grimsdale, W. Straßer. VIII, 276 pages, 124 figs., 1991

Advances in Computer Graphics VI. Images: Synthesis, Analysis, and Interaction. Edited by G. Garcia, I. Herman. IX, 449 pages, 186 figs., 1991

Intelligent CAD Systems III. Practical Experience and Evaluation. Edited by P. J. W. ten Hagen, P. J. Veerkamp. X, 270 pages, 116 figs., 1991

Graphics and Communications. Edited by D. B. Arnold, R. A. Day, D. A. Duce, C. Fuhrhop, J. R. Gallop, R. Maybury, D. C. Sutcliffe. VIII, 274 pages, 84 figs., 1991

Photorealism in Computer Graphics. Edited by K. Bouatouch, C. Bouville. XVI, 230 pages, 118 figs., 1992

Advances in Computer Graphics Hardware V. Rendering, Ray Tracing and Visualization Systems. Edited by R. L. Grimsdale, A. Kaufman. VIII, 174 pages, 97 figs., 1992

Multimedia. Systems, Interaction and Applications. Edited by L. Kjelldahl. VIII, 355 pages, 129 figs., 1992. Out of print

Advances in Scientific Visualization. Edited by F. H. Post, A. J. S. Hin. X, 212 pages, 141 figs., 47 in color, 1992

Computer Graphics and Mathematics. Edited by B. Falcidieno, I. Herman, C. Pienovi. VII, 318 pages, 159 figs., 8 in color, 1992

Rendering, Visualization and Rasterization Hardware. Edited by A. Kaufman. VIII, 196 pages, 100 figs., 1993

Visualization in Scientific Computing. Edited by M. Grave, Y. Le Lous, W. T. Hewitt. XI, 218 pages, 120 figs., 1994

Photorealistic Rendering in Computer Graphics. Edited by P. Brunet, F. W. Jansen. X, 286 pages, 175 figs., 1994

From Object Modelling to Advanced Visual Communication. Edited by S. Coquillart, W. Straßer, P. Stucki. VII, 305 pages, 128 figs., 38 in color, 1994

MIX
Papier aus verantwortungsvollen Quellen
Paper from responsible sources
FSC® C105338

If you have any concerns about our products,
you can contact us on
ProductSafety@springernature.com

In case Publisher is established outside the EU,
the EU authorized representative is:
Springer Nature Customer Service Center GmbH
Europaplatz 3, 69115 Heidelberg, Germany

Printed by Libri Plureos GmbH
in Hamburg, Germany